# SpringerBriefs in Evolutionary Biology

W0037563

For further volumes:
http://www.springer.com/series/10207

SpringerBriefs in Evolutionary Biology

Rituparna Bose · Alexander J. Bartholomew

# Macroevolution in Deep Time

With Foreword by Dr. Ashok Sahni

 Springer

Rituparna Bose
The City University of New York
Bayside, NY
USA

Alexander J. Bartholomew
State University of New York
New Paltz, NY
USA

ISSN 2192-8134
ISBN 978-1-4614-6475-4
DOI 10.1007/978-1-4614-6476-1
Springer New York Heidelberg Dordrecht London

ISSN 2192-8142    (electronic)
ISBN 978-1-4614-6476-1    (eBook)

Library of Congress Control Number: 2013930356

Printed on acid-free paper

Springer is part of Springer Science+Business Media (www.springer.com)

**Parts of this monograph have been published in the Springer book in**
*Paleoenvironmental Sciences*:

Bose, R., 2012, Biodiversity and Evolutionary Ecology of Extinct Organisms.
Springer ISBN 978-3-642-31720-0, 100 p.

# Foreword

Science thrives on anomalies, controversies, and competing hypotheses. In the nineteenth century, Charles Darwin awakened the interest of the common man by discovering a coherent theory of evolution that included a principle and a plausible explanation and soon faced considerable opposition from people holding traditional viewpoints. He established the principle of descent with modification, and demonstrated that the principle can be accomplished by 'natural selection'. Darwin believed in gradualism, i.e., new forms of life typically come into being via slow and continuous processes. He believed 'gradualistic evolution' to be inherently more rational and scientific than an evolution theory that advocated "saltation", a process in which species evolve by jumps and starts.

After over a century of Darwin's findings, scientists Eldredge and Gould published a landmark paper on development of the evolutionary theory on 'punctuated equilibrium'. Their theory suggested that it is stasis that dominates the history of most fossil species accompanied by rapid change during speciation and extinction. Thereafter, controversies were raised between the two established evolutionary theories—phyletic gradualism and punctuated equilibrium. Researchers, since then, have been testing these theories using both extinct and living species in the past four decades. Rituparna Bose has tested the two theories using well preserved, extinct fossil brachiopod species from the Middle Devonian strata of Michigan Basin.

Bose describes evolutionary mechanisms in extinct brachiopods to understand their response to environmental change. She has applied novel approaches in solving complex hypotheses in popular aspects of evolutionary biology, such as the theories of phyletic gradualism and punctuated equilibrium. She has statistically determined the extent to which these evolutionary theories are dominant in extinct invertebrates. Such studies have implications in predicting future biodiversity, extinction, ecosystem conservation, and climate change.

I recommend this book to the student community at large and especially to advanced professionals in the field of evolutionary biology, palaeobiology, and geosciences. A detailed bibliography at the end of the book will allow readers to access the primary literature if required.

Dr. Ashok Sahni
Professor Emeritus, Centre of Advanced Study in Geology
Panjab University, India
President, The Palaeontological Society of India
Fellow of the Geological Society of India
Fellow of the Indian National Science Academy
Fellow of the Academy of Sciences for the Developing World (FTWAS)
Associe Etranger, Geological Societe, France
Fellow of the Humboldt Foundation, Bonn

# Preface

Organisms are morphologically complex and thus, studying their morphological response to environmental change over time and space can shed light on the tempo and modes of evolution. Identifying the various environmental parameters in deep time and additionally, determining the precise modes of evolution in extinct organisms still remains one of the greatest challenges in Earth's history.

In this book, evolutionary mechanisms in extinct organisms were described to understand how organisms evolved in deep time. Novel approaches were used in solving complex hypotheses in popular aspects of evolutionary biology, such as the theories of phyletic gradualism and punctuated equilibrium. A statistical analysis was carried out to see which of these evolutionary theories are dominant in extinct invertebrates. Such studies have implications in predicting future biodiversity, ecosystem conservation, and climate change. Chapter 1 discusses the history of evolutionary theories and progressive advancements in this field of study. Chapter 2 discusses the major findings of evolutionary patterns observed in a Devonian brachiopod fossil lineage. Chapter 3 discusses the comparative morphologies observed in fossil brachiopod lineages across correlated geological strata in the Devonian time period and gauges the scope of environmental change in terms of local or regional aspect. Chapter 4 discusses the implications of such studies in evolutionary biology and paleontology in both extinct and living forms in the context of future perspectives.

This book will be a valuable read to evolutionary biologists, paleontologists, ecologists, geologists, environmental, and climate scientists. It may be used in undergraduate classes but will certainly help post-graduate students and advanced professionals.

The first author is grateful to Prof. David Polly at Indiana University, Bloomington for his valuable suggestions. The foreword for the book has been written by Prof. Ashok Sahni (Elected Fellow of Indian National Science Academy and National Geoscience Lifetime achievement Awardee for Excellence-2009).

# Acknowledgments

This research was funded by the Theodore Roosevelt Memorial Research Grant received from the American Museum of Natural History. The first author is thankful to Daniel Miller for providing access to the Michigan Museum of Paleontology collections. The first author would like to express her sincere gratitude towards David Polly for his kind suggestions. A special thanks to Lindsey Leighton for his suggestions with sample collection.

# Contents

# Chapter 1
# Advancement in Evolutionary Theories

**Abstract** The mode and tempo of morphological change in lineages over geologic time has been a hotly debated topic in paleontology and biology over the last few decades. Fossil lineages have long been tested in the light of evolutionary models—punctuated equilibrium and phyletic gradualism and yet, we lack Darwin's vision of an integrated understanding of evolution. Darwin, well recognized as the father of evolution, had been successful in awakening the human minds by discovering a coherent theory of evolution. His theory included a principle and a plausible explanation and was based on the principle of descent with modification, accomplished by natural selection. His ideas revolved around a central theme of gradualist evolution. However, soon his opponents advocated another theory in evolution called the punctuated equilibrium. The later theory revolved around the key concept that evolution happens in a stepwise fashion. Thus, studying evolution in the light of these models can possibly help with understanding the mechanism behind the process.

## 1.1 History of Evolution

In the eighteenth century, a few naturalists introduced the idea that life may have been changing since creation. By the end of the seventeenth century, paleontologists had discovered and investigated a rich treasure of fossil collections from Europe, suggesting that life has been undergoing through some change since the past. And in 1801, a French naturalist named Jean Baptiste Lamarck proposed a conceptual theory of evolution. Lamarck, born in 1744, started his scientific career as a botanist, but later in his career, he became one of the founding professors of invertebrate zoology in the National Museum of Natural History in France. He became a pioneer of invertebrates, and his work on classifying worms, spiders, molluscs, and other invertebrates gained great importance far ahead of his time. Lamarck believed that organisms had to change their behavioral response in order to cope with the changing environment. It was he who believed that long necks of giraffes evolved as generations of giraffes reached even higher to feed on leaves of tall trees. Lamarck also proposed that organisms were driven from simple to increasingly more complex forms. He was one of the true pioneers who had recognized that evolution did occur in nature, but his ideas were sadly abandoned by other naturalists of his times who were

R. Bose and A. J. Bartholomew, *Macroevolution in Deep Time*, SpringerBriefs in Evolutionary Biology, DOI: 10.1007/978-1-4614-6476-1_1, © The Author(s) 2013

driven by the ideas of forces of creation and could not see what Lamarck was trying to explain. Lamarck died in 1829 but his ideas and innovations were given recognition by other naturalists and writers thereafter and finally, *The Origin of Species* was published by Charles Darwin in 1859. Charles Darwin was the first English naturalist to discover a coherent theory of evolution in 1859 that included a principle and a plausible explanation. He established the principle of descent with modification, and also that the principle can be accomplished by 'natural selection', where the best trait adapts to the new suitable environment. His ideas soon faced considerable opposition from people who possessed traditional opinions. Darwin believed in gradualism, which states that during the course of evolution, new forms of life typically come into being via slow and continuous processes. He believed 'gradualistic evolution' to be inherently more rational and scientific than saltational accounts of evolution. However, his theory did not take into account the geological history.

In many ways, however, Darwin's central argument is very different from Lamarck's. Darwin believed that complex organisms evolved as a consequence of adapting to its surrounding local environmental conditions. His ideas revolved around the fact that the history of life is not driven by complexity alone but it is evolutionary change in these complex forms from one generation to the next that gives rise to new forms, solely via natural selection. He also argued that species could also go extinct rather than just changing into new forms.

While Darwin's ideas thrived on different concepts, like Lamarck, he also relied on much the same evidence for evolution (such as vestigial structures and artificial selection through breeding). Darwin mistakenly accepted the fact that changes acquired during an organism's lifetime could be passed on to its offspring. Due to a lack of understanding of the concepts of heredity among scientists, Lamarckian inheritance remained popular throughout the 1800s. But with the discovery of genes, his theory was finally abandoned for the most part. But Lamarck, whom Darwin described as "this justly celebrated naturalist", remains a major figure in the history of biology for envisioning evolutionary change for the first time. After over a century of Darwin's explanations on gradualist evolutionary theory, American paleontologists, Niles Eldredge and Stephen Jay Gould published a landmark paper on development of the evolutionary theory on 'punctuated equilibrium'. Their theory suggested that it is stasis that dominates the history of most fossil species accompanied by rapid change during speciation and extinction. There still continues to remain controversies on the two evolutionary theories (phyletic gradualism and punctuated equilibrium), and thus, researchers have been extensively testing these theories since the past four decades.

### 1.1.1 Punctuated Equilibrium

The theory of 'Punctuated Equilibrium' attempts to explain the macroevolutionary role of species and speciation as expressed in geologic time (Gould 2007). This theory makes the strong claim that in most cases, effectively no change accumulates at all. A species at its last appearance before extinction, does not differ systematically from the anatomy of its initial entry into the fossil record, usually several million years before. In other words, punctuated equilibrium is a theory in evolutionary biology which states that most sexually reproducing species experience little change for most of their geological history, and that when phenotypic evolution does occur, it is localized in rare, rapid events of branching speciation (called cladogenesis).

Its statements about rapidity and stability describe the history of individual species. As a central proposition, 'Punctuated Equilibrium' model holds that the great majority of species, as evidenced by their anatomical and geographical histories in the fossil record, originate in geological moments (punctuations) and then persist in stasis throughout their long durations. Sepkoski (1997) gives a low estimate of 4 my for the average duration of fossil species; mean values vary widely across groups and times, with terrestrial vertebrates at lesser durations and most marine invertebrates in the higher ranges. In any case, geological longevity achieves its primary measure in millions of years, not thousands (Sepkoski 1997). Of course, fluctuation of mean values through time is well recognized. Criteria need to be set for permissible fluctuation in average values through time. Two issues that need to be resolved are: (1) the amount of allowable difference between beginning and ending samples of a species and (2) the range of permissible fluctuation through time. These issues mainly deal with the evolutionary hypothesis that little or no change accumulates by phyletic change during the history of most species. Since no statistical right exists to expect that the last samples will be identical with the first, one should predict that (i) the final samples shall not differ statistically, by some conventionally chosen criterion, from the initial forms; and at the very least that (ii) the final samples shall not lie generally outside the range of fluctuation observed during the history of the species. The observations from these tested hypotheses allow geologists to read the history of individual species in a more accurate sense.

## 1.1.2 Who Coined the Punctuated Equilibrium Theory?

Niles Elredge and Stephen Jay Gould coined the theory of evolution that occurs in steps. In 1972, paleontologists Niles Eldredge and Stephen Jay Gould (graduate students at the American Museum of Natural History, AMNH) published a landmark paper developing this novel idea. Their paper was built upon Ernst Mayr's theory of geographic speciation, Michael Lerner's theories of developmental and genetic homeostasis, as well as their own empirical research. Eldredge and Gould proposed that the degree of gradualism championed by Charles Darwin was virtually nonexistent in the fossil record, and that stasis dominates the history of most fossil species (Eldredge and Gould 1972).

Eldredge and Gould were the first to point out that modern speciation theory would not predict gradual transitions over millions of years, but instead the sudden

appearance of new species in the fossil record punctuated by long periods of species stability, or equilibrium is expected. Eldredge and Gould not only showed that
paleontologists had been not in line with biologists for decades, but also that they
had unconsciously been trying to fit the fossil record into the gradualistic mode of
evolution. The few supposed examples of gradual evolution were featured in the
journals and textbooks, but paleontologists had long been silent about their "dirty
little trade secret" which suggests that most species appear suddenly in the fossil
record and show no appreciable change for millions of years until their extinction.

### 1.1.3  Proponents and Opponents of the Punctuated Equilibrium Model

One of the main proponents of gradualism, Philip Gingerich (1976, 1980a, b,
1987), showed just two or three examples of supposed gradual evolution in early
Eocene (about 50–55 million years old) mammals from the Bighorn Basin of
northwestern Wyoming. But a detailed examination of the entire mammal fauna
(monographed by Bown 1979 and studied by Gingerich 1989) shows that most
of the rest of the species do not change gradually through time. Later, Gingerich
(2001) himself illustrates stasis for longer timescales in Cenozoic mammals by
considering the shape of a heuristic time-form evolutionary lattice.

"The Eldredge-Gould concept of punctuated equilibria" gained wide accept-
ance among paleontologists after the 1970s. It attempts to account for the following paradox: 'Within continuously sampled lineages, one rarely finds the
gradual morphological trends predicted by Darwinian evolution; rather, change
occurs with the sudden appearance of new, well-differentiated species'. Eldredge
and Gould equate such appearances with speciation, although the details of these
events are not preserved. The punctuated equilibrium model has been widely
accepted, not because it has a compelling theoretical basis but because it appears
to resolve a dilemma that occupied several human minds for several years. Apart
from the obvious sampling problems inherent to the observations that stimulated
the PE model, and apart from its intrinsic circularity (one could argue that speciation can occur only when phyletic change is rapid, not vice versa), the model is
more of an ad hoc explanation than theory, and it rests on shaky ground (Ricklefs
1978). This is from a review of the book "Patterns of Evolution as Illustrated by
the Fossil Record" edited by Hallam (1977), and the section that has been replaced
by ellipses is as follows: They suggest that change occurs rapidly, by geologic
standards, in small peripheral populations. They believe that evolution is acceler-
ated in such populations because they contain a small random sample of the gene
pool of the parent population (founder effect) and therefore can diverge rapidly
just by chance and because they can respond to local selection pressure that may
differ from those encountered by the parent population. Eventually some of these
divergent, peripheral populations are favored by changed environmental conditions
(species selection) and so they increase and spread rapidly into fossil assemblages.

So Ricklefs disagrees with the idea of Punctuated Equilibria. But does he disagree with the idea of evolution? Apparently, he did not agree with the idea of evolution. He believed that the fossil record clearly is inadequate for many purposes and to be absolutely certain, one can discern general trends in morphology and diversity within phylogenetic groups only. However, most paleontologists agreed to the fact that adaptive radiations occur in brief bursts at the family to class levels of taxonomy, often after the decline of ecologically related groups, and are followed by long periods of evolutionary stability.

Some studies where invertebrate fossil species lineages were tested for mode of evolution show a direct link between morphology and changes in environment. Sheldon (1987) proposed parallel gradualistic evolution in eight trilobite species lineages over a period of 3 my based on a study of ~15,000 Ordovician trilobites from Central Wales. A number of pygidial ribs were analyzed over seven sections, further subdivided into 400 sampling localities (mean stratigraphic thickness of 23 cm) in the Teretiusculus Shales (Llandeilo Series) from the Builth inlier. These shales accumulated in persistent, low energy, dysaerobic conditions, probably in a silled marine basin several hundred meters deep representing a narrowly fluctuating, less oxygenated, and stable environment. Perhaps, he mentioned that this kind of gradual phyletic evolution can only be sustained by organisms living in or able to track narrowly fluctuating, slowly changing environments, whereas stasis, almost paradoxically tends to prevail in more widely fluctuating, rapidly changing environments. Finally he proposed the Plus ca change model in 1996 which proposes that over geologic timescales (e.g. million years), gradualism is characteristic of narrowly fluctuating, relatively stable environments such as the terrestrial tropics and the deep sea. By contrast net stasis with occasional punctuations is expected to prevail in more unstable environments that dominate the fossil record, especially shallow seas (Sheldon 1996).

The patterns we observe in biological communities and evolutionary radiations are the sum of many lower order processes and interactions. And even though Ricklefs disagrees with Punctuated Equilibria, he does not discount it completely. Even though Eldredge and Gould may be proven right, their model and other recent models in paleontology, should not be accorded the status of a major synthesis. Investigations need to be carried out continuously and the revised models of evolution should be compared with the older models to make sense of the theory of evolution in large.

### 1.1.4 Speculations: If PE Is True, Then What Caused Species to Remain Stable for Such a Long Period of Time Despite Environmental Changes?

If species are static through millions of years in spite of environmental changes, there must be some sort of homeostatic mechanism that preserves this stability beyond what traditional reductionist Neo-Darwinism once postulated (Prothero 1992). Mayr (1992) argues that it is merely the integration of species as complex wholes, so that small-scale changes are insufficient to upset the complex balance of integrated genes. Others suggest that fundamental developmental constraints play an important role in restricting the possible avenues of change (Gould and Lewontin 1979; Kauffman 1983). Still others suggest that there might be characteristics of species that may not have been discovered yet by geneticists and evolutionary biologists, characteristics which operate on scales of millions of generations and years (Vrba 1980; Vrba and Eldredge 1984).

## 1.2 Some Competing Ideas on How Evolution Works

Punctuated equilibrium is commonly contrasted against the theory of phyletic gradualism, which states that evolution generally occurs uniformly and by the steady and gradual transformation of whole lineages (anagenesis) (Fig. 1.1). In this view, evolution is seen as generally smooth and continuous. A few characteristic differences in the two theories are listed in Table 1.1.

Phyletic gradualism is a macroevolutionary hypothesis rooted in uniformitarianism. The hypothesis states that species continue to adapt to new challenges over the course of their history, gradually becoming new species. Gradualism holds that every individual is the same species as its parents, and that there is no clear line of demarcation between the old species and the new species. It holds that the species is not a fixed type, and that the population, not the individual, evolves. During this process, evolution occurs at a slow and smooth (but not necessarily constant) rate, even on a geological timescale (cf. punctuated equilibrium).

**Fig. 1.1** Two models of evolution. The two theories of evolution showing contrasting patterns in time

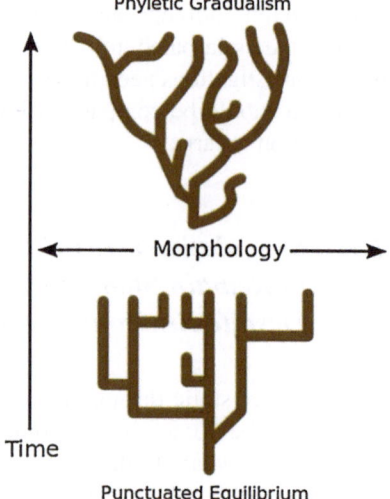

**Table 1.1** Contrasting theories of evolution

|  | Phyletic gradualism | Punctuated equilibrium |
|---|---|---|
| Distribution of rate of phenotypic change | Uniformly low; no rate increase with splitting | Episodic; high during speciation; low to absent within species |
| Direction of phenotypic change within a species | Unidirectional; evolution is phyletic (Phyletic evolution or anagenesis involves the gradual transformation of one species to another) | Oscillates about a stable mean |
| Population size required for adaptive change to occur | Small or large; it can occur in whole species | Only small isolated populations |
| Can species-specific character change occur in stable environments, i.e., purely as a function of time | Yes (evolutionary trend is mostly in the direction of specialization of species-specific character); it is speeded up by environmental change | No; it requires environmental change |
| How can new species arise? | By phyletic speciation in sympatry; and by allopatric speciation in small or large populations | Generally only by allopatric speciation in small isolated populations |
| Implication for species | Species may be arbitrary subdivisions of a lineage continuum | Species are real, discrete entities, with beginnings and terminations (defn by Mayr: real discrete entities that interbreed with individuals within them and are reproductively isolated from other species) |

# 1.3  Gap in the Fossil Record

Charles Darwin believed that evolution was a slow and gradual process. He did not believe this process to be "perfectly smooth," but rather thought it to be "stepwise", with a species evolving and accumulating small variations over long periods of time. Darwin assumed that if evolution is gradual then there should be a record in fossils of small incremental change within a species. But in many cases, Darwin was unable to find most of these intermediate forms. He blamed and related the lack of transitional forms to the gaps in the fossil record, because the chances of each of those critical changing forms having been preserved as fossils are very small. Darwin's conclusive observation was that while the broad outline of the fossil history of life was consistent with descent with modification and natural selection, the geological record was too incomplete and too poorly known to document in detail the transformation of species (Hunt 2010).

Later in 1972, evolutionary scientists Stephen Jay Gould and Niles Eldredge proposed another explanation for the numerous gaps in the fossil record. They suggested that the "gaps" were real, representing periods of stasis in morphology. They termed this mode of evolution "punctuated equilibrium". This means that species are generally morphologically stable, changing little for millions of years. This leisurely pace is "punctuated" by a rapid burst of change that results in a new species. According to this idea, changes leading to a new species do not usually occur from slow incremental change in the mainstream population of a species, but occur in those populations living in the periphery, or in small geographically isolated populations where their gene pools vary more widely due to the slightly different environmental conditions where they dwell. When the environment changes, these "peripheral" or "geographic isolates" possess variation in morphology which might enable them to have an adaptive advantage, leading to greater reproductive success. These new successful morphotypes spread through the geographic range of the ancestral species, appearing as a new morphology where once the older forms were present (Eldredge and Gould 1972).

Today, the fossil record is much resolved, and its strengths and weaknesses are much better understood Hunt and Chapman (2001). As Darwin envisioned, under the most promising circumstances, it is possible to document the transformation of a lineage by natural selection in fossil strata. There is also a good quantitative record of evolutionary patterns in fossil lineages over typical paleontological resolutions (1,000–10,000,000 years). At these scales, phenotypic evolution within lineages appears to be overwhelmingly non-directional and often surprisingly slow. The fluctuating trajectories captured in the fossil record are not inconsistent with the central theme of natural selection as an evolutionary mechanism, but they probably would not have been predicted without the benefit of an empirical fossil record Hunt and Chapman (2001).

## 1.4  Processes Behind Evolution

One of the well-known processes of evolution is mutation of genes leading to variation and reproduction (Fig. 1.2).

Natural selection is a process by which genetic mutations is known to become more common in successive generations of a population resulting in more variation in their reproductive offspring (Darwin 1859). The central concept of natural selection is the adaptation ability and evolutionary fitness of an organism (Orr 2009). This measures the organism's genetic contribution to the next generation. However, this is not the same as the total number of offsprings: instead fitness refers to the proportion of subsequent generations that carry an organism's genes (Haldane 1959). Consequently, if an allele (one of two or more forms of a gene) increases fitness more than the other alleles of that gene, then with each generation this allele will become more common within the population. These traits are said to be "selected for". Examples of traits that can increase fitness are enhanced survival and increased fecundity. Conversely, the lower fitness caused by having a less beneficial or deleterious allele results in this allele becoming rarer—they are "selected against" (Lande and Arnold 1983). It is important to understand that the

**Fig. 1.2** Mutation and selection leading to reproduction where mutation followed by natural selection, results in a population with darker coloration

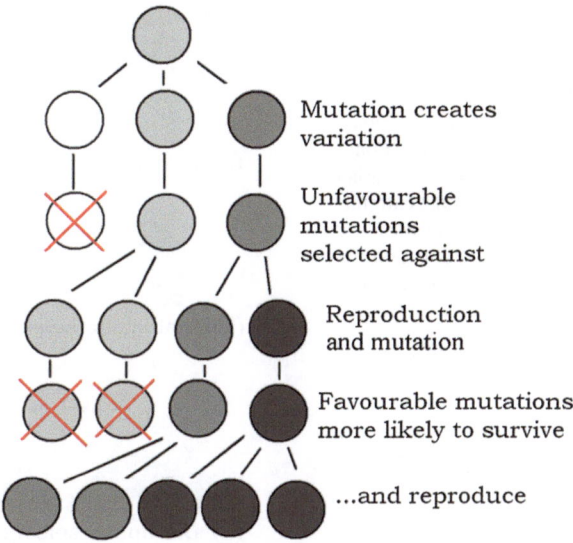

Mutation creates variation

Unfavourable mutations selected against

Reproduction and mutation

Favourable mutations more likely to survive

...and reproduce

fitness of an allele is not a fixed characteristic; if the environment changes, previously neutral or harmful traits may become beneficial and previously beneficial traits become harmful (Futuyma 2005).

Natural selection within a population for a trait like 'height' can vary across a range of values and can be categorized into three different types, also known as 'modes of evolution' (Fig. 1.3). The first is directional selection, which is described as a shift in the average value of a trait over time—for example organisms slowly getting taller (Hoekstra et al. 2001). Secondly, disruptive selection is selection for extreme trait values and often results in two different values becoming most common, with selection against the average value. This would be when either short or tall organisms had an advantage over medium height. Finally, in stabilizing selection there is selection against extreme trait values on both ends, which causes a decrease in variance around the average value, thus leading to less diversity (Hurst 2009). This would, for example, cause organisms to slowly become all the same height.

In population genetics, there are different modes or mechanisms of natural selection (Andrews 2010). Directional selection occurs when natural selection favors a single phenotype and therefore allele frequency continuously shifts in one direction (Fig. 1.3). In directional selection, the advantageous allele will increase in frequency independently of its dominance relative to other alleles (i.e., even if the advantageous allele is recessive, it will eventually become fixed). Disruptive selection, also called diversifying selection, refers to changes in population genetics that simultaneously favor individuals at both extremes of the distribution. When disruptive selection operates, individuals at the extremes contribute more offspring than those in the center, producing two peaks in the distribution of a particular trait (Fig. 1.3). Stabilizing selection is a type of natural selection in

**Fig. 1.3** Types (modes) of natural selection. (**a**) Directional selection: favors individuals at one end of the phenotypic selection; (**b**) Stabilizing selection: favors intermediate variants by acting against extreme phenotypes; and (**c**) Diversifying selection: favors extreme over intermediate phenotypes

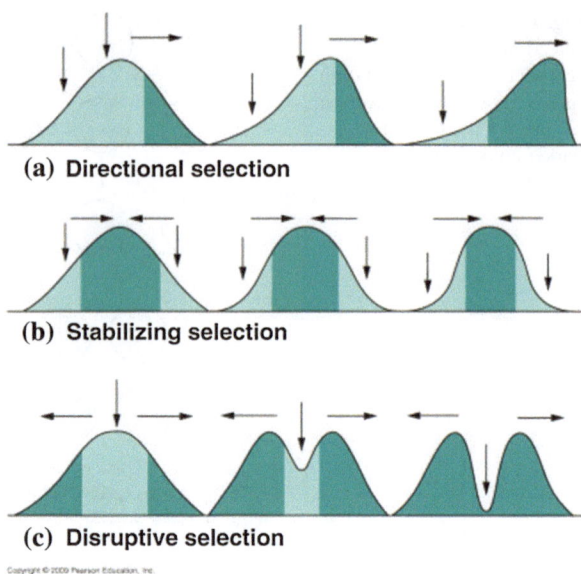

**(a) Directional selection**

**(b) Stabilizing selection**

**(c) Disruptive selection**

which genetic diversity decreases as the population stabilizes on a particular trait value. Alternatively, it can be referred to the extreme values of the character that are selected against. This is probably the most common mechanism of action for natural selection. Stabilizing selection operates most of the time in most populations and this type of selection acts to prevent divergence of form and function. In this way, the anatomy of some organisms, such as sharks and ferns, has remained largely unchanged for millions of years (Andrews 2010).

The different types of evolutionary modes can be represented by computer-generated simulation graphs (Polly 2004). These graphs show evolution of complex morphologies over geological timescales of millions of generations (Fig. 1.4). Each of the different modes of selection leave a distinctive imprint on the distribution of morphological distances, a phenomenon that is well known for univariate traits (e.g., Gingerich 1993; Roopnarine 2001; Polly 2004).

## 1.5 Tempo and Mode of Evolution

Evolution has been rarely quantified before the 1990s. Gingerich (1993) used a unit called Haldanes that can be easily interpreted in terms of quantitative evolutionary genetics. He used log rate versus log interval (LRI) distribution to determine modes and rates of evolution along timescales. His LRI plot provides estimates of the average intrinsic generation to generation rate of an evolutionary time series, which is significant in understanding the long-term rates of evolution. Intrinsic rates refer to standard deviations per generation on a timescale of

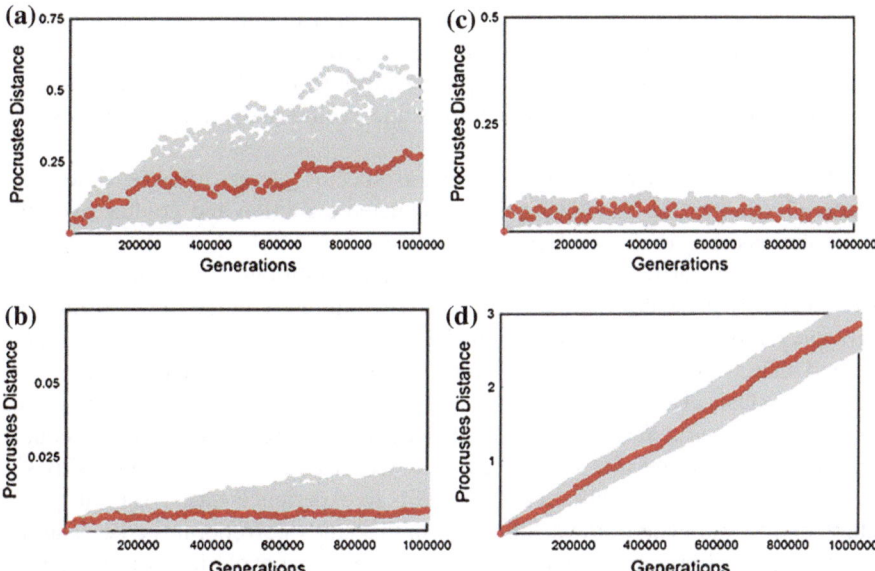

**Fig. 1.4** Modes of evolution of molar tooth morphology in shrews, simulated under four different evolutionary modes—**a** Randomly fluctuating selection. **b** Random drift (Ne = 70). **c** Stabilizing selection. **d** Directionally biased selection; Y-axis represents morphological distance and X-axis represents geological timescales in millions of generations (from Polly 2004)

one generation. Evolutionary rates in invertebrate and vertebrate fossil lineages can be intrinsic similar to modern lineages, however, it is extrinsic environmental factors that gauge evolutionary change or stasis (Gingerich 1993). Thus, both intrinsic (change from any step to next step) and net rates (change from any step to any step) need proper scrutiny to determine the actual rates of evolution in organisms. It was Haldane (1949) who first proposed that evolutionary rates can be calculated in terms of time that is measured in generation (and not years); generations in time represent reproductive cycle which is different in different organisms.

One haldane is defined as change by a factor of one standard deviation per generation. Phenotypic standard deviations are natural measures of normally distributed variation for traits under study. Use of haldane units is advantageous over other units of evolutionary rates proposed in the past (Darwin and Simpson), as standard deviations are natural measures of variation and generations in organisms are natural counts of their reproductive cycles and thus, rates calculated in standard deviation per generation are both dimensionless and dimension-independent (Gingerich 1993). Haldane evolutionary rate is calculated as follows:

$$\text{Rate(h)} = \frac{\left(\ln x_2 / S_{\ln x}\right) - \left(\ln x_1 / S_{\ln x}\right)}{(t_2 - t_1)} = \frac{z_2 - z_1}{(t_2 - t_1)} \tag{1.1}$$

where ln $x_1$ and ln $x_2$ are sample means of ln measurements at times $t_1$ and $t_2$. Ln measurements of $x_1$s and $x_2$s are pooled together as $s_{lnx}$ which is the pooled standard deviation for respective samples. Gingerich (1993) further used simulation plots under random walk models to gauge evolutionary rates in species lineages (Fig. 1.5).

After testing rates of change of several species along different timescales, Gingerich came to the conclusion that punctuated evolution and gradual evolution are evolutions measured on different timescales.

Later in the early twenty-first century, a paleobiologist named Gene Hunt performed a large-scale, statistical survey of evolutionary mode in fossil lineages. He analyzed over 250 sequences of evolving traits using the maximum-likelihood method that were fit to three evolutionary models: directional change, random walk, and stasis. He noted that only 5 % of fossil sequences reveal directional evolution and the remaining 95 % of sequences were divided nearly equally between random walks and stasis. He also noted that variables related to body size were significantly less likely than shape traits to experience stasis. The rarity with which directional evolution was observed in this study corroborates a key claim of punctuated equilibria and suggests that truly directional evolution is infrequent or, perhaps more importantly, of short enough duration so as to rarely register in paleontological sampling (Hunt 2007).

Hunt used likelihood-based framework for fitting and comparing models of phyletic evolution. Because of its usefulness and historical importance, he focuses on a general form of the random walk model. The long-term dynamics of this

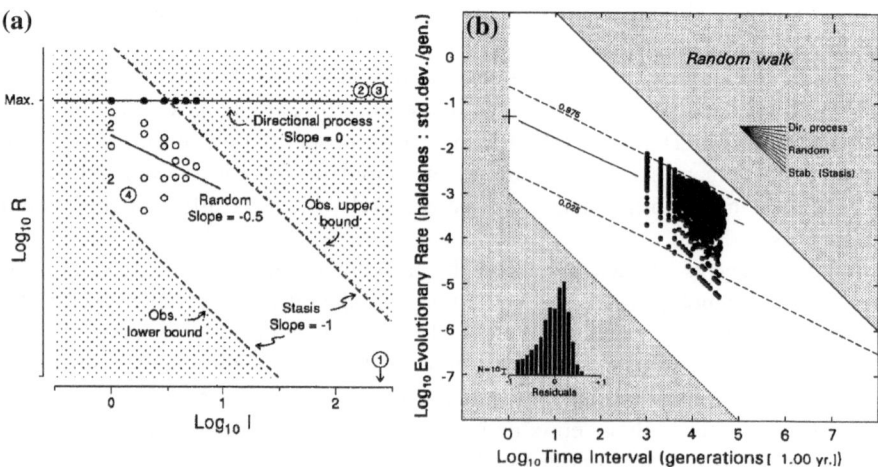

**Fig. 1.5** Rates of change in relation to time and measurement interval represented by an LRI distribution graph ($Log_{10}R$ vs. $Log_{10}I$ is change over time). **a** A slope value of $-1$ represents stasis ($-1$), $-0.5$ represents random, and 0.0 represents directional change; **b** a simulation plot tested against a random walk model, shows several data points that represent random mode of evolution (after Gingerich 1993)

model depend on just two parameters: the mean ($\mu_s$) and variance ($\sigma_s^2$) of the distribution of evolutionary transitions (or "steps").

The value of $\mu_s$ governs the directionality of trait evolution, whereas $\sigma_s^2$ determines the volatility of evolutionary changes around the directional trend (Hunt 2006). Simulations show that these two parameters can be inferred reliably from paleontological data regardless of how completely the evolving lineage is sampled. In addition to random walk models, suitable modification of the likelihood function permits consideration of a wide range of alternative evolutionary models.

Hunt (2007) found that for any dataset, the three evolutionary models—directional change, unbiased random walk, and stasis could be fit by maximum-likelihood method using the software package paleoTS in the freely available statistical programming environment R (R core team 2007). Model support that balances the goodness of fit (log-likelihood) with model complexity (the number of model parameters) was assessed by using the bias-corrected Akaike Information Criterion (AIC$_C$) (Anderson et al. 2000). An important advantage of this model-based approach is that no evolutionary mode is granted privileged null status; all models are compared on an equal footing based on their empirical support (Hunt 2007).

Toward the end of the first decade of the twenty-first century, maximum-likelihood method gained popularity in understanding evolutionary mode of species lineages in the fossil record (Polly 2008; Bose 2012a). Evolutionary rate and mode in morphological divergence were assessed using the maximum-likelihood method (Polly 2008). This method uses the following equation to estimate rate and mode simultaneously, where D is morphological divergence (Procrustes distance), r is the mean rate of morphological divergence, t is divergence time, and a is a coefficient that ranges from 0 to 1, where 0 represents complete stabilizing selection (stasis), 0.5 represents perfect random divergence (Brownian motion), and 1 represents perfect diversifying (directional) selection (Polly 2008) (Fig. 1.6).

$$D = \mathrm{rt}^\wedge a \tag{1.2}$$

This method is used to find the parameters '$r$' and '$a$' which maximize the likelihood of the data and are thus the best estimates for rate and mode (Polly 2008; Bose 2012a). This method estimates the mean per step evolutionary rates, and the degree of stabilizing or diversifying selection from a matrix of pairwise morphological distances and divergence times (Polly 2008). Morphological divergence ($D$) is generally calculated as pairwise Procrustes distances among samples. Divergence time ($t$) is an estimate of the total time in millions of years that the two samples have been diverging independently (Fig. 1.6).

# 1.6  A Case Study on Brachiopods

This study was the first attempt to apply geometric morphometrics to study complex morphology of atrypid brachiopod species lineages over 5 my timescale. The goal of this study was to determine patterns of change in morphological shape in the context

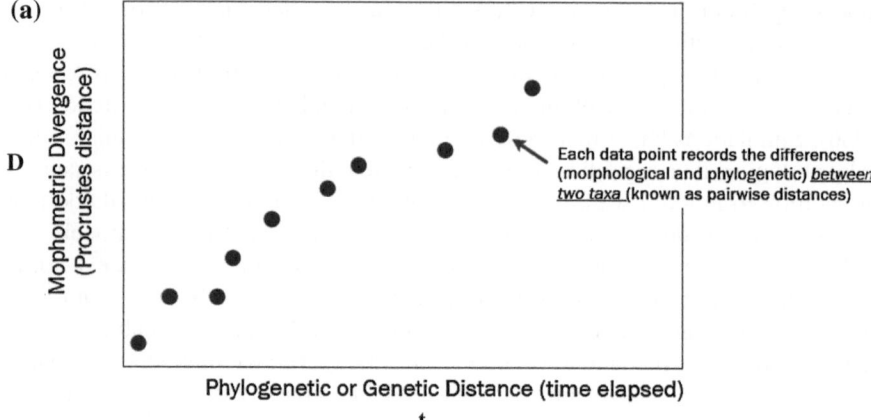

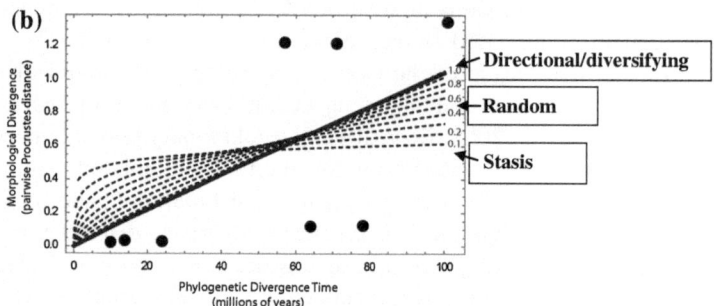

**Fig. 1.6** (**a**) Plot of morphological divergence against phylogenetic, genetic, or geographic distance (Polly 2012); (**b**) Graph showing morphometric divergence (pairwise Procrustes distances) and phylogenetic divergence (millions of years). The series of *dashed lines* show the expected relationship between morphological and phylogenetic divergence time from strong stabilizing selection (0.1), through random divergence (0.5), to diversifying (directional) selection (1.0). The maximum-likelihood estimate of this relationship, shown by the dark line, suggests that these samples have experienced diversifying selection (Bose 2012a)

of evolutionary theory. Previous studies in brachiopod morphometrics have proven to be effective (Tyler and Leighton 2011; Bose et al. 2011; Bose 2012a, 2012b). Thus, this study, where well preserved samples were tested against both the theories of evolution, will serve as a great database for future evolutionary biologists.

## References

Anderson DRK, Burnham P, Thompson WL (2000) Null hypothesis testing: problems, prevalence and an alternative. J Wildl Manag 64:912–923
Andrews CA (2010) Natural selection, genetic drift, and gene flow do not act in isolation in natural populations. Nat Education Knowl 1(10):5

Bose R (2012a) A new morphometric model in distinguishing two closely related extinct brachiopod species. Hist Biol Int J Paleobiol. doi: 10.1080/08912963.2012.658568

Bose R (2012b) Biodiversity and evolutionary ecology of extinct organisms. Earth system sciences. Springer, New York

Bose R, Schneider CL, Leighton LR, Polly PD (2011) Influence of atrypid morphological shape on Devonian episkeletobiont assemblages from the lower Genshaw formation of the traverse group of Michigan: a geometric morphometric approach. Palaeogeogr Palaeoclimatol Palaeoecol 310(3–4):427–441

Bown TM (1979) Geology and mammalian paleontology of the Sand Creek facies, lower Willwood formation (lower Eocene). Washakie County, Wyoming, Geol Surv Wyoming Memoir 2:1–151

Darwin C (1859) On the origin of species by means of natural selection, or the preservation of favoured races in the struggle for life. John Murray, London, p 502

Eldredge N, Gould SJ (1972) Punctuated equilibria: an alternative to phyletic gradualism. In: Schopf TJM (ed) Models in paleobiology. Freeman, San Francisco, pp 82–115

Futuyma DJ (2005) Evolution. Sinauer Associates, Inc. ISBN 0-87893-187-2, Sunderland, Massachusetts

Gingerich PD (1980a) Evolutionary patterns in early Cenozoic mammals. Annu Rev Earth Planet Sci 8:407–424

Gingerich PD (1980b) Early cenozoic paleontology and stratigraphy of the bighorn basin, Wyoming: 1880–1980. Univ Michigan Papers Paleontol 24:1–146

Gingerich PD (1987) Early Eocene bats (Mammalia, Chiroptera) and other vertebrates in freshwater limestones of the Willwood Formation, Clarks Fork Basin, Wyoming. Contributions Museum Paleontol Univ Michigan 27:275–320

Gingerich PD (1989) Paleocene and early Eocene of the Bighorn and Clark's Fork Basins, Wyoming. In: Flynn JJ (ed) 28th international geological congress field trip guidebook T322. American Geophysical Union, Washington, pp 47–57

Gingerich PD (1993) Quantification and comparison of evolutionary rates. Am J Sci 293A:453–478

Gingerich PD (2001) Rates of evolution on the time scale of the evolutionary process. In: Hendry AP, Kinnison MT (eds) Contemporary microevolution: rate, pattern, and process, Kluwer Academic Publishers, Dordrecht, Genetica 112/113:127–144

Gingerich PD, Rose KD (1976) Partial skull of the plesiadapiform primate Ignacius from the early Eocene of Wyoming. Contributions Museum Paleontol Univ Michigan 24:181–189

Gould SJ (2007) Punctuated equilibrium. Belknap of Harvard University Press, Cambridge, p 408

Gould SJ, Lewontin RC (1979) The spandrels of san marco and the panglossian paradigm: a critique of the adaptationist programme. Proc R Soc Lond B 205:581–598

Haldane JBS (1949) Suggestions as to quantitative measurement of rates of evolution. Evolution 3:51–56

Haldane JBS (1959) The theory of natural selection today. Nature 183:710–713

Hallam A (1977) Developments in palaeontology and stratigraphy 5. Elsevier, Amsterdam, p 591

Hoekstra H, Hoekstra J, Berrigan D, Vignieri S, Hoang A, Hill C, Beerli P, Kingsolver J (2001) Strength and tempo of directional selection in the wild. Proc Natl Acad Sci USA 98(16):9157–9160

Hunt G, Chapman RE (2001) Evaluating hypotheses of instar-grouping in arthropods: a maximum likelihood approach. Paleobiol 27(3):466–484

Hunt G (2006) Fitting and comparing models of phyletic evolution: random walks and beyond. Palaeobiology 32:578–601

Hunt G (2007) The relative importance of directional change, random walks, and stasis in the evolution of fossil lineages. Proc Natl Acad Sci 104:18404–18408

Hunt G (2010) Evolution in fossil lineages: paleontology and the origin of species. Am Nat 176:S61–S76

Hurst LD (2009) Fundamental concepts in genetics: genetics and the understanding of selection. Nat Rev Genet 10:83–93

Kauffman SA (1983) Developmental constraints: internal factors in evolution. In: Goodwin BC, Holder N, Wylie CC (eds) Development and evolution. Cambridge University Press, Cambridge, pp 195–225

Lande R, Arnold SJ (1983) The measurement of selection on correlated characters. Evolution 37:1210–1226

Mayr E (1992) Speciational evolution or punctuated equilibria. In: Somit A, Peterson SA (eds) The dynamics of evolution. Cornell University Press, Ithaca, pp 21–53

Orr HA (2009) Fitness and its role in evolutionary genetics. Nat Rev Genet 10:531–539

Polly PD (2004) On the simulation of the evolution of morphological shape: multivariate shape under selection and drift. Palaeontologia Electronica 7:1–28

Polly PD (2008) Adaptive zones and the pinniped ankle: a 3D quantitative analysis of carnivoran tarsal evolution. In: Sargis E, Dagosto M (eds) Mammalian evolutionary morphology: a tribute to Frederick S. Szalay, Springer, Dordrecht, pp 165–194

Polly PD (2012) Phenotypic evolution and phylogenetic comparative methods. Geometric morphometrics course. Indiana University, Bloomington

Prothero DR (1992) Punctuated. Equilibrium. At twenty: a paleontological perspective. Skeptic 1:38–47

R Development Core Team, authors. R (2007) A language and environment for statistical computing. Available at www.R-project.org

Ricklefs RE (1978) Paleontologists confronting macroevolution. Science 199:59

Roopnarine PD (2001) The description and classification of evolutionary mode: a computational approach. Paleobiology 27:446–465

Sepkoski JJ (1997) Biodiversity: past, present, and future. J Paleontol 71:533–539

Sheldon PR (1987) Parallel gradualistic evolution of Ordovician trilobites. Nature 330:561–563

Sheldon PR (1996) Plus ça change—a model for stasis and evolution in different environments. Palaeogeogr Palaeoclimatol Palaeoecol 127:209–227

Tyler CL, Leighton LR (2011) Detecting competition in the fossil record: support for character displacement among Ordovician Brachiopods. Palaeogeogr Palaeoclimatol Palaeoecol 307:205–217

Vrba ES (1980) Evolution, species, and fossils: how does life evolve? S Afr J Sci 76:61–84

Vrba ES, Eldredge N (1984) Individuals, hierarchies and processes; towards a more complete evolutionary theory. Paleobiology 10:146–171

# Chapter 2
# Evolutionary Change: A Case Study of Extinct Brachiopod Species

**Abstract** Brachiopods serve as great tools for quantifying evolution of early life. In this study, morphological characters were quantified for 1,100 individual specimens of a single atrypid species from the Middle Devonian Traverse Group of Michigan using geometric morphometric methods. Seven landmark measurements were taken on dorsal valve, ventral valve, and anterior and posterior regions. Specimens were partitioned by their occurrence in four stratigraphic horizons (Bell Shale, Ferron Point, Genshaw Formation, and Norway Point) from the Traverse Group of northeastern Michigan outcrop. Multivariate statistical analyses were performed to test patterns and processes of morphological shape change of species over 6 million year interval of time. Maximum-likelihood method was used to determine the evolutionary rate and mode in morphological divergence in this species over time. If punctuated equilibrium model holds true for brachiopods, then one would expect no significant differences between samples of the species *Pseudoatrypa cf. lineata* from successive stratigraphic units of the 6 million year Middle Devonian Traverse Group over time. Multivariate analysis shows significant shape differences between different time horizons ($p \leq 0.01$) with considerable overlap in morphology excepting abrupt deviation in morphology in the uppermost occurrence. Maximum-likelihood tests further confirm near stasis to near random divergence mode of evolution with slow to moderate rates of evolution for this species lineage. In contrast, if the species evolved in a gradual, directional manner, then one would expect samples close together in time to be more similar to one another than those more separated in time. Euclidean based cluster analysis shows samples closely spaced in time are more similar than those that are widely separated in time. While these results appear to partially support the gradualistic model hypothesis, morphological trend from principal component scores shows substantial morphological overlap among the three lower successions (Bell Shale, Ferron Point, and Genshaw Formation) with some deviation from the uppermost succession (Norway Point) suggesting major influence of stasis and punctuation. Thus, while stasis may have been predominant, evident from stable morphologies observed in the lower strata, anagenetic evolution may also have played an important role as evident from the abrupt change observed within this species later in time. Thus, slow to moderate rates of evolution in this species lineage with stable morphologies in the lower three strata supports stasis, but abrupt

R. Bose and A. J. Bartholomew, *Macroevolution in Deep Time*, SpringerBriefs
in Evolutionary Biology, DOI: 10.1007/978-1-4614-6476-1_2,
© The Author(s) 2013

change in the uppermost strata supports gradual anagenetic evolution within the species later in time. Overall, the morphometric data for the *P. cf. lineata* species lineage are consistent with the stratigraphic succession of Traverse Group.

## 2.1 Introduction

The mode and rate of morphological change in evolutionary fossil and extant species lineages over geologic time has been a hotly debated topic in paleontology and biology over the last three decades. Phyletic gradualism model proposed by several researchers (Simpson 1953; Mayr 1963) and punctuated equilibrium model proposed by others (Eldredge and Gould 1972; Gould and Eldredge 1977; Stanley 1979) have largely attributed to the prevalence of stasis in species over long periods of time. While stasis has been tested in taxa from Middle Devonian Hamilton group, no one has studied the Traverse Group taxa for testing morphological patterns.

The relative importance of stasis has been studied for many fossil lineages. Shape traits like bivalve convexity studied for Neogene *Chesapecten nefrens* for over 4 million year have yielded results in support for stasis (Hunt 2007). Gingerich (2001) illustrates stasis for longer time scales in Cenozoic mammals by considering the shape of a heuristic time-form evolutionary lattice. Roopnarine (2001) originally interpreted the Mio-Pliocene foraminifers *Globorotalia plesiotumida-tumida* as the punctuated anagenetic transformation of *Globorotalia plesiotumida* to *G. tumida* but using the iterative method of subseries of an original stratophenetic series (by measuring the presence of deviations from statistical randomness as the lineage evolves), he interpreted this as "constrained stasis". Polly (2001) used the Brownian motion model of evolution for reconstructing ancestral nodes versus observed nodes in a fully resolved phylogeny of fossil carnivorans and found that change is constrained in molar areas over longer time intervals. He found that there is change of the sort that one would expect by random selection (Brownian motion) when patterns are measured on a small scale, but that at a very broad scale, (e.g., all mammals) the amount of change was less than if one extrapolated the small scale change to the large scale, suggesting that there is constraint on divergence.

Work of this sort has also been done specifically on brachiopods. Previous workers (Lieberman et al. 1995) tested the morphological variability of the common brachiopod species lineages from size measurements on the pedicle valves of 401 *Mediospirifer audaculus* and 614 *Athyris spiriferoides* from successive stratigraphic horizons in the Hamilton Group (5 million year) of New York. They found morphological overlap within these species between the lowermost and uppermost strata with some variations in the intervening samples of the Hamilton Group. Goldman and Mitchell (1990) tested the internal morphology of three brachiopod species of the Hamilton Group of western New York from size measurements and found only one species of Late Givetian age showed some species level change. Isaacson and Perry (1977) have not found any significant change in *Tropidoleptus carinatus* of the Givetian age Hamilton Group from the lowest to its highest occurrence, spanning some 40 million year and further, Eldredge tested the same fauna using morphometrics and found almost no significant morphological change in this unit (Brett and Baird 1995). Overall, a majority of Hamilton brachiopod species lineages represent stasis in most cases while some minor evolutionary changes have been recorded in some cases. Unidirectional evolution over geological time intervals is highly unlikely and evolutionary reversals are common (Sheldon 1996).

Not all workers agree that stasis is common within species (Gingerich 1976, 1985; Sheldon 1987; Geary 1995; Webber and Hunda 2007; Geary et al. 2010). Two studies that have most prominently challenged stasis are Gingerich's (1976) work on fossil mammals from the Bighorn Basin of the western United States and Sheldon's (1987) (Fig. 3.6) work on Ordovician trilobites from Wales. Sheldon (1987) reported evidence of phyletic gradualism based on a study of eight lineages of 15,000 Ordovician trilobites from central Wales over a 3 million year interval. He believed that subdividing a species lineage into subspecies often gives a false impression of punctuation and stasis. In another study of Cenozoic gastropod species lineages from Pannonian Basin system, Geary (1995) has emphasized the importance of gradual change in evolutionary paleontological studies. Webber and Hunda (2007) using geometric morphometrics found that certain aspects of the morphological shape of the Upper Ordovician trilobite *Flexicalymene granulosa* change with varying paleoenvironmental conditions during the deposition of Kope and lower Fairview Formations spanning over a 2 million year interval.

This study tests hypothesis of stasis in conjunction with morphological patterns observed both in a gradualistic and punctuated equilibrium framework. In this study, morphological shape change pattern over time is assessed in *Pseudoatrypa cf. lineata* species lineage from the Givetian 6 million year Traverse Group strata. This brachiopod species *Pseudoatrypa cf. lineata* (Webster 1921), was subjected to geometric morphometric and multivariate statistical analyses to examine mode and rate of morphological shape evolution. This species was sampled from the three lower strata of the Middle Devonian Traverse Group of Eastern North America and one upper strata near the top of the Traverse Group. Whether they are present in almost all of the stratigraphic units of the Traverse Group or not, remains unknown. However, plentiful of this species is recognized from the Bell Shale, Ferron Point, Genshaw Formation, and Norway Point stratigraphic units from Alpena and Presque Isle counties of the northeastern outcrop of Michigan.

The Traverse fauna, and the strata of the Traverse Group, have been the subject of many detailed stratigraphic and paleoenvironmental studies (e.g., Ehlers and Kesling 1970; Kesling et al. 1974; Wylie and Huntoon 2003), and these studies allow fossil specimens collected from the Traverse Group to be placed in a paleoenvironment setting. This intensely studied geological system of the Traverse Group presents a good opportunity to study the relationship between morphological change and environment over time in an individual species lineage *Pseudoatrypa cf. lineata*. For this species, the amount of morphological change through time was determined within shale lithologic settings (nearshore environment), to characterize morphological changes in a more confined paleoenvironmental setting over extended stratigraphic intervals. Whether morphological shape change corresponds with water depth data of Wylie and Huntoon (2003) or not, is also tested in this study.

### 2.1.1 Hypotheses

(1) If this species *P. cf. lineata* evolved according to the punctuated equilibrium model, in which morphological change occurs predominantly at speciation, thus, remaining static in most of its life, then we would expect no significant differences between samples of the species from successive stratigraphic units of 6 million year Middle Devonian Traverse Group over time; (2) If the species evolved in a gradual, directional manner, then we would expect samples close together in time to be more similar to one another than those more separated in time. Besides, one would also expect directional selection to be acting on this species.

## 2.2 Geological Background

The geologic setting used to test the proposed hypotheses in the *P. cf. lineata* species lineage is the Traverse Group, a package of rocks from Michigan State, that spans roughly 6.5 million year of the Middle Devonian and Lower Upper Devonian. Their appearance in North America seems to have been driven by the post-Eifelian augmentation of the Acadian orogeny (Wylie and Huntoon 2003). The Traverse Group and its fauna are associated with the influx of siliciclastic sedimentation from this Orogeny (Brett 1986; Cooper et al. 1942; Ettensohn 1985; Ehlers and Kesling 1970; Wylie and Huntoon 2003).

The richly fossiliferous strata from the 6 million year (380.0–374.0 million year) Middle Devonian Traverse Group rocks (Wylie and Huntoon 2003) of the northeastern outcrop of Michigan from Alpena and Presque Isle counties are used for this study. Among the 11 Traverse Group formations ((from base upward): Bell Shale, Rockport Quarry Limestone, Ferron Point Formation, Genshaw Formation, Newton Creek Limestone, Alpena Limestone, Four Mile Dam Formation, Norway Point Formation, Potter Farm Formation, Thunder Bay Limestone, and Squaw Bay Limestone), four of the formations are included in this study for collecting *P. cf. lineata* samples. The study interval for this investigation, which includes the stratigraphic ranges in the northeastern Michigan—Bell Shale, Ferron Point, Genshaw Formation, and Norway Point, was deposited approximately 380 million years ago and represents more than 6.0 million years duration. The Givetian age Traverse

Group is equivalent to the Hamilton Group of New York (Ehlers and Kesling 1970). Thus, this case study will represent a comparative analysis of brachiopod morphological shape change patterns after Lieberman et al. 1995.

These strata comprise a nearly 565-ft. thick succession of sedimentary rocks, primarily shales, claystones, and limestones, which were deposited in predominantly supratidal to nearshore marine settings (Ehlers and Kesling 1970; Wylie and Huntoon 2003). The Bell Shale, about 68 feet in thickness, consists of a basal crinoid rich lag and shales, which were deposited with water depth ranging from 82 to 147 ft. The Ferron Point, about 42 ft. in thickness, consists of soft shales and limestones, deposited with water depth approximately 131 ft. The Genshaw Formation, 116.5 ft. in thickness, consists of soft shales and argillaceous limestones, with water depth ranging from 82 to 131 ft. Norway Point Formation, 45 ft. in thickness, consists of abundant shales and claystones, with limestones, deposited at approximately, 6.5-ft. water depth (Wylie and Huntoon 2003). The formations chosen for data collection in this study are dominated by shales (Wylie and Huntoon 2003) and thus for this study, sampling restricted to shale beds in the four formations allows morphological analysis in a more or less stable environmental setting. Our data were collected from thinly bedded shales of Bell Shale, Ferron Point, Lower Genshaw, and Norway Point formations.

Samples used in morphological shape study are from the Michigan Museum of Paleontology Collections. Some of the samples come from the collections of Alex Bartholomew from State University of New York that have now been deposited at the Indiana University Paleontology Collections.

## 2.3 Materials and Methods

Samples for this study were collected from Bell Shale, Ferron Point, Genshaw, and Norway Point stratigraphic intervals of the Middle Devonian Traverse Group (6 million year) from the northeastern outcrop of Michigan from Alpena and Presque Isle counties. First, atrypid samples were qualitatively examined and identified based on external morphological characteristics. 1,124 specimens of *P. cf. lineata* was used from a total of four different shale beds at six localities of four above-mentioned strata (Bell Shale = 131; Ferron Point = 330; Genshaw = 506; Norway Point = 157) in Michigan. Samples used in morphological shape study are from the Michigan Museum of Paleontology Collections. Some of the collections of Alex Bartholomew from State University of New York have also been used for this study which are now deposited at the Indiana University Paleontology Collections.

For this study, material has been confined to shale lithology representing supratidal to nearshore environments from 0 to 50 m water depth which may have been interrupted by occasional storms. Overall, these habitats represent low energy conditions interrupted by occasionally very high energy conditions, with normal ranges of marine salinity, oxygen, and temperature. Consequently, the sampled brachiopods to be analyzed will be less subject to taphonomic distortion through transport or mechanical destruction, while still retaining sufficient diversity and abundance to provide representative samples. However, depending upon the frequency of storm events and the turbidity of the water column from influx of siliciclastic sedimentation from the Taghanic Onlap, the faunas in these settings can be influenced by a variety of environmental parameters like light intensity variations, sedimentation rate, dissolved oxygen concentration, salinity, and temperature variations.

*Pseudoatrypa cf. lineata* (Brachiopoda) makes an ideal candidate for quantitatively analyzing the morphological shape evolution after controlling for the environmental setting. This species displays a suitable amount of morphological complexity for geometric morphometrics as landmarks selected on various points of the shell help fully describe the shell's morphology. Besides, this species ranges through some of the Traverse group formations (Bell Shale, Ferron Point, Genshaw, and Norway Point) such that it provides enough stratigraphic coverage to recognize potential temporal trends in morphology (Fig. 2.1). Finally, abundant samples are available for a statistical significant analysis across time.

In this study, morphological analyses have focused on seven landmark points (Fig. 2.2), each landmark point representing the same location on each specimen to capture the biologically most meaningful shape. These landmark points represent discrete juxtapositions, functional equivalents, and extremal points (Table 2.1) (Bookstein 1991; MacLeod and Forey 2002; Zelditch et al. 2004).

Though these landmark points are not biologically homologous, they correspond among diverse forms (sensu Bookstein 1991), which is appropriate for

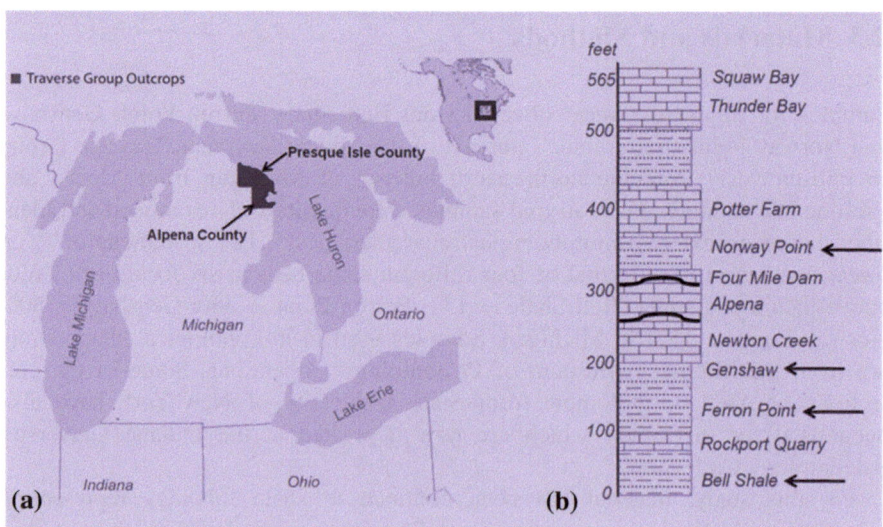

**Fig. 2.1 a** Study area map showing sample field site collection locations in the Michigan State. **b** Stratigraphy of the Traverse Group of Michigan. *Arrows* indicate sampled strata from where brachiopod fossils were collected

**Table 2.1** Landmark points on the *Pseudoatrypa cf. lineata* shell representing geometric positions that are biologically functional

| Landmark points | Area of the shell | Landmark descriptions |
|---|---|---|
| 1 | D | Tip of umbo |
| 2 | D/V | Junction on the hinge of dorsal valve interarea, ventral valve interarea, and commissure |
| 3 | D/V | Midpoint of specimen length projected onto commissure, length midpoint based on length of baseline |
| 4 | D/V | Extreme edge of anterior commissure adjacent to L5 |
| 5 | D/V | Edge of commissure perpendicular to hinge, in line with L1 (on sulcate specimens, this point coincided with the lowest point of the sulcus) |
| 6 | D | Maximum height of curvature |
| 7 | V | Lowest point of interarea/on the pedicle foramen |

*D* dorsal and *V* ventral

analyses attempting to capture morphological shape changes (Rohlf and Marcus 1993) or any functional aspects of the species. Brachiopods are bilaterally symmetrical organisms and each side is a mirror image of the other, i.e., each half captures the shape of the organism. Thus, for all individuals, measurements were taken on half of the specimen (dorsal view right, ventral view left, and one side of anterior and posterior views) (Fig. 2.2).

Data were captured using Thin Plate Spline dig (TPSdig) software for digitizing landmarks for geometric morphometrics. Procrustes analysis (Rohlf 1990, 1999; Rohlf and Slice 1990; Slice 2001) was performed on original shape data, rotating, translating, and scaling all landmarks to remove all size effects, while maintaining their geometric relationships (Procrustes superimposition). The Procrustes coordinates were used in all morphometric analyses. Principal component analysis was performed to determine the morphological variation between samples from four stratigraphic intervals (Bell Shale, Ferron Point, Genshaw, and Norway Point) along their major principal component axes (1 and 2) in the shape morphospace. Mean morphological shape trend was constructed from principal component scores for both dorsal and ventral valves along the stratigraphic units of the Traverse Group (Fig. 2.3). Thin Plate Spline (Bookstein 1989) visualization plots were created to detect morphological shifts in individual landmark points over time. Multivariate analysis of variance (MANOVA) was performed to test shape variation between these samples (Fig. 2.4).

Euclidean based cluster analysis was performed to determine Procrustes pairwise distances from mean morphological shape of samples between the Traverse Group stratigraphic intervals (Table 2.2, Fig. 2.5). Due to the lack of appropriate absolute age dates, time matrix was constructed by calculating time in million

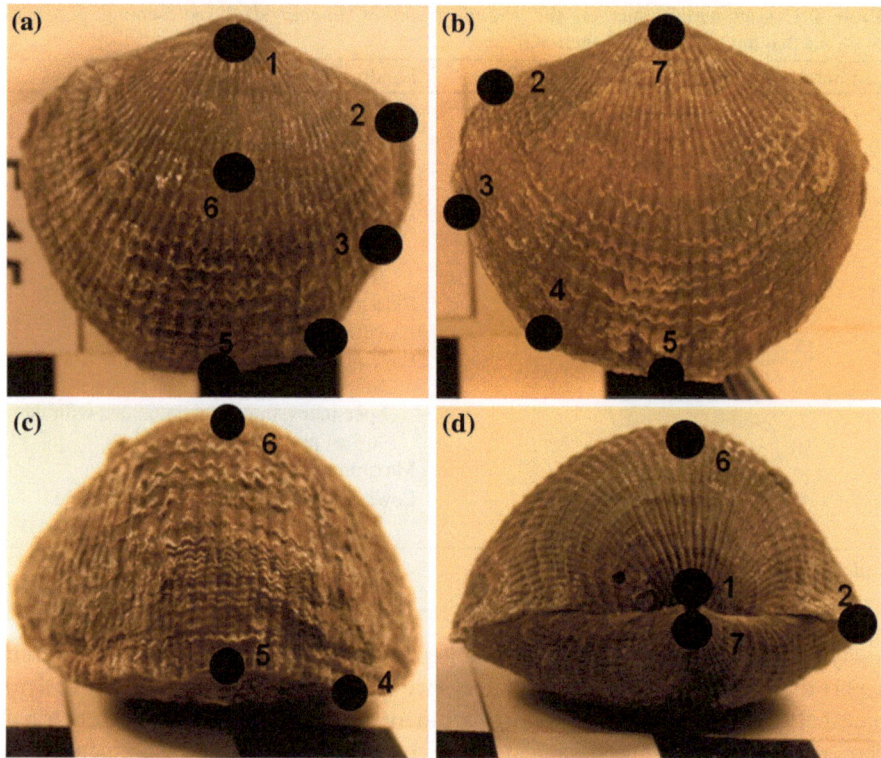

**Fig. 2.2** Seven landmark points on *Pseudoatrypa cf. lineata* shells with distribution of **a** six landmark points on right side of dorsal valve, **b** five landmark points on left side of ventral valve, **c** three landmark points on right side of anterior margin area of shell, and **d** four landmark points on right side of posterior hinge view of shell

years among each of the four stratigraphic intervals of the Traverse Group based on the stratigraphic thicknesses of the formations obtained from Wylie and Huntoon (2003) (Tables 2.3, 2.4). Thus, the morphological distances and time data accumulated from morphometric and stratigraphic differences were used to construct time distance matrix.

Evolutionary rate and mode in morphological divergence was assessed using the maximum-likelihood method of Polly (2008). This method estimates the mean per-step evolutionary rate and the degree of stabilizing or diversifying selection from a matrix of pairwise morphological distances and divergence times. Morphological distance was calculated as pairwise Procrustes distances among *P. cf. lineata* taxon and divergence time was calculated from the stratigraphic thickness converted to millions of years on Wiley and Huntoon's (2003) stratigraphic column, which is an estimate of the total time in millions of years between strata that the species has been diverging independently. The method uses the following equation to estimate rate and mode simultaneously,

**Fig. 2.3** Mean morphological
shape trend in **a** dorsal and
**b** ventral valves of *P. cf.
lineata* along four Traverse
Group formations (Bell
Shale, Ferron Point, Genshaw
Formation and Norway Point);
**c** morphological variation
in dorsal valves along the
four units representing
percent variation along
principal component axes 1
and 2 (PC1 = 36.02 % and
PC2 = 21.62 %)

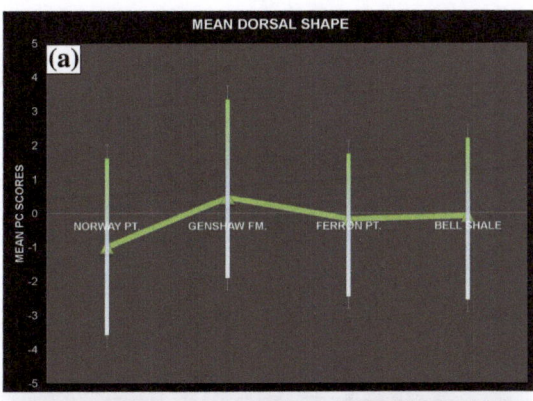

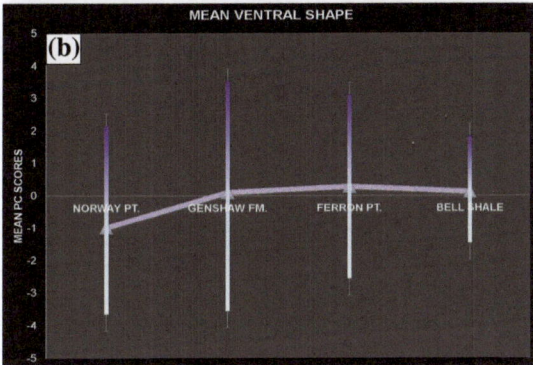

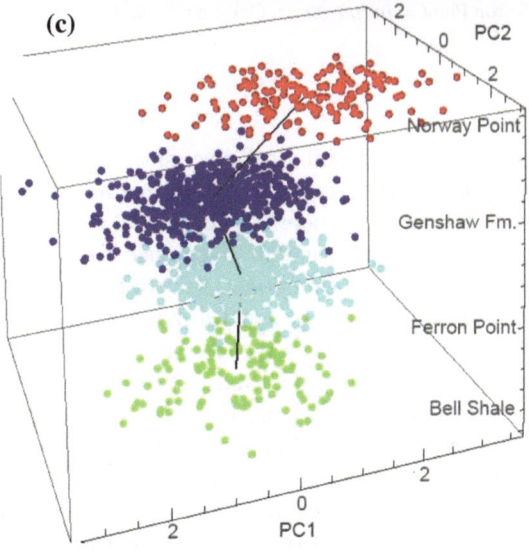

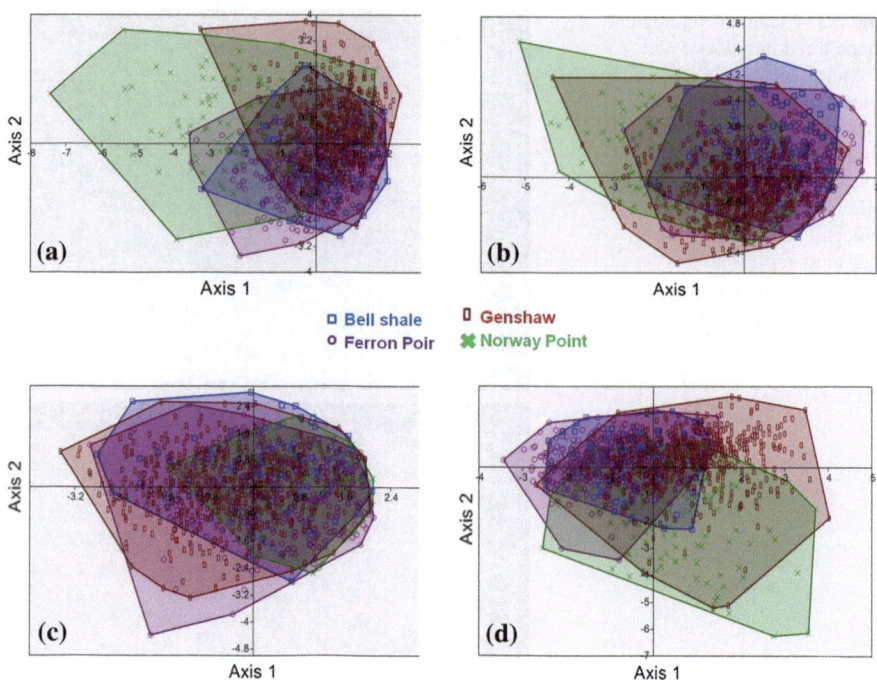

**Fig. 2.4** Canonical Variate Analysis showing morphometric differences between samples from Bell Shale, Ferron Point, Genshaw Formation, and Norway Point ($p < 0.01$) in **a** dorsal valves, **b** ventral valves, **c** anterior margin area, and **d** posterior hinge area (*Note* Bell Shale = *blue square*, Ferron Point = *purple circle*, Genshaw = *dark red rectangle*, Norway Point = *green cross*)

$$D = rt\,^{\wedge}\, a \tag{2.1}$$

where D is morphological divergence (Procrustes distance), r is the mean rate of morphological divergence, t is divergence time, and a is a coefficient that ranges from 0.0 to 1.0, where 0.0 represents complete stabilizing selection (stasis), 0.5 represents perfect random divergence (Brownian motion), and 1.0 represents perfect diversifying (directional) selection (Polly 2008). Maximum-likelihood is used to find the parameters r and a that maximize the likelihood of the data, and are thus the best estimates for rate and mode (Fig. 2.6). The data were bootstrapped 1,000 times to generate standard errors for these estimates. This method is derived directly from the work presented by Polly (2004) and is mathematically related to other methods in evolutionary genetics (Felsenstein 1988; Gingerich 1993; Hunt 2007; Lande 1976; Roopnarine 2003).

Morphological patterns in *Pseudoatrypa cf. lineata* are compared with water depth data from Wylie and Huntoon (2003). The mean morphological PC1 scores for the samples are statistically correlated with water depth using linear regression analysis.

**Table 2.2** Procrustes pairwise distances for *Pseudoatrypa cf. lineata* lineage among stratigraphic units in time for (a) dorsal valve, (b) ventral valve, (c) anterior marginal area, and (d) posterior hinge area

|            | Bell    | Ferron  | Genshaw | Norway  |
|------------|---------|---------|---------|---------|
| *Dorsal*   |         |         |         |         |
| Bell       | 0       | 0.94315 | 1.1128  | 1.9706  |
| Ferron     | 0.94315 | 0       | 1.3173  | 1.8075  |
| Genshaw    | 1.1128  | 1.3173  | 0       | 1.8232  |
| Norway     | 1.9706  | 1.8075  | 1.8232  | 0       |
| *Ventral*  |         |         |         |         |
| Bell       | 0       | 0.84557 | 1.1987  | 1.686   |
| Ferron     | 0.84557 | 0       | 0.79457 | 1.7012  |
| Genshaw    | 1.1987  | 0.79457 | 0       | 1.295   |
| Norway     | 1.686   | 1.7012  | 1.295   | 0       |
| *Anterior* |         |         |         |         |
| Bell       | 0       | 0.63057 | 0.75156 | 0.5672  |
| Ferron     | 0.63057 | 0       | 0.85516 | 0.47276 |
| Genshaw    | 0.75156 | 0.85516 | 0       | 1.1393  |
| Norway     | 0.5672  | 0.47276 | 1.1393  | 0       |
| *Posterior*|         |         |         |         |
| Bell       | 0       | 0.74795 | 1.1225  | 1.7207  |
| Ferron     | 0.74795 | 0       | 1.5494  | 1.8043  |
| Genshaw    | 1.1225  | 1.5494  | 0       | 1.6468  |
| Norway     | 1.7207  | 1.8043  | 1.6468  | 0       |

## 2.4  Results

Among the atrypid samples, *Pseudoatrypa cf. lineata* was recognized from four Traverse Group formations based on qualitative traits. Main characteristics used for identification included medium- to large-size shells with globose dorsibiconvex-convexiplanar shells and an inflated hemispherical dome-like dorsal valve, shell length exceeding width slightly in all growth stages, subquadrate shell outline, broad to angular fold developed posterior of mid-valve, becoming more pronounced toward anterior margin in large adult shells (30 mm), exterior of both valves with fine radial tubular ribs (9–10/5 mm at anterior margin), regularly spaced concentric lamellae crowding toward anterior and lateral margins in larger adults (20 mm length), and short frills rarely preserved.

Eleven hundred and twenty four specimens of *P. cf. lineata* were analyzed for morphological shape change pattern over time. Geometric morphometric analysis was used to test the morphological differences in shell shape between the Traverse Group stratigraphic intervals. The major principal component axis, axis 1, explained for 36.02 % variation in dorsal valves, 28.80 % variation in ventral valves, 62.59 % variation in anterior margin area, and 82.87 % variation in posterior hinge area. The second major principal component axis, axis 2, explained for 21.62 % variation in dorsal valves, 23.85 % variation in ventral valves, 35.83 % variation in anterior margin area, and 13.07 % variation in posterior hinge area. Principal component

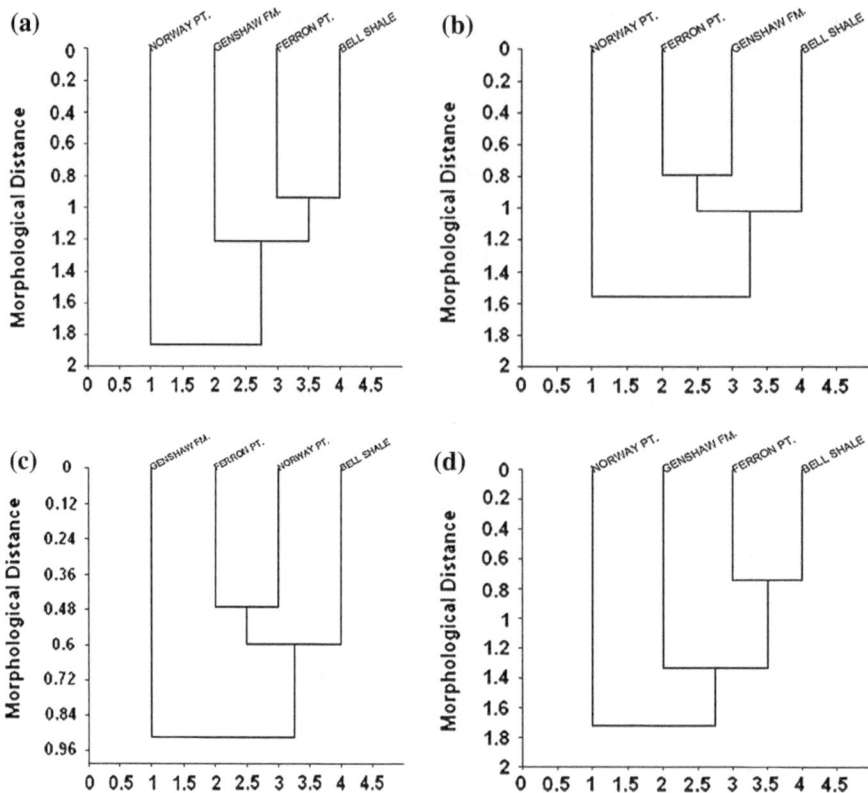

**Fig. 2.5** Morphological links for *P. cf. lineata* in Traverse group formations of **a** dorsal valves, **b** ventral valves, **c** anterior margin area, and **d** posterior hinge area

**Table 2.3** Eleven stratigraphic units in the Traverse Group representing various thicknesses in feet and time in million years

|  | Stratigraphic units | Thickness (m) | Time (million year) |
| --- | --- | --- | --- |
| Traverse group | Squaw Bay | 2.85 | 0.10 |
|  | Thunder Bay | 4.8 | 0.17 |
|  | Potter Farm | 30.6 | 1.08 |
|  | **Norway** | **13.5** | **0.48** |
|  | Four Mile Dam | 6.3 | 0.22 |
|  | Alpena | 23.7 | 0.84 |
|  | Newton Creek | 7.5 | 0.27 |
|  | **Genshaw** | **34.95** | **1.24** |
|  | Ferron | 12.6 | 0.45 |
|  | Rockport | 12.6 | 0.45 |
|  | **Bell** | **20.4** | **0.72** |
|  | **Total thickness** | **169.5** | **6.02** |

Time in million years for each unit calculated from thickness estimates of each unit w.r.t the total thickness of the Traverse Group (169.5 m). *Pseudoatrypa cf. lineata* samples represent the stratigraphic units in bold

**Table 2.4** Time matrix (million years) calculated for the four stratigraphic units used for this study from thicknesses of the strata in time

|         | Bell | Ferron | Genshaw | Norway |
|---------|------|--------|---------|--------|
| Bell    | 0    | 1.16   | 1.61    | 4.18   |
| Ferron  | 1.16 | 0      | 0.45    | 3.02   |
| Genshaw | 1.61 | 0.45   | 0       | 2.57   |
| Norway  | 4.18 | 3.02   | 2.57    | 0      |

analysis indicated that there is considerable shape variation within samples from each stratigraphic interval and that the lower three stratigraphic horizons overlap considerably in the morphology of both valves with some deviation in Norway Point samples (Fig. 2.3a–c). Thin Plate Spline visualization plots show the mean morphological shape of these samples from the four stratigraphic intervals are quite similar, which along with the overlap in morphological variation, demonstrates the samples from the four strata are not substantially different in morphological shape.

MANOVA, however, detects a small but significant statistical difference in mean shape ($p < 0.01$) between the stratigraphic intervals (Fig. 2.4) based on Hotelling's p values. The statistical significance of the MANOVA demonstrates that there is some real differentiation between the samples from the four stratigraphic horizons in shell shape, but the substantial overlap in variation and the difficulty in visually distinguishing the differences in shell shape suggests to us that all these samples of *P. cf. lineata* species show little morphological change over time (Figs. 2.3, 2.4). However, abrupt deviation in mean morphological shape of the Norway Point samples (Fig. 2.3a–c) suggests gives some evidence of morphological change in this species later in time.

Procrustes distances, which are the units of difference in the principal components space, between the mean shape of the brachiopod shell, ranged from 0.47 to 1.97 Procrustes units for the stratigraphic intervals of the Traverse Group (Fig. 2.5, Table 2.2). Closely spaced stratigraphic units represent smaller Procrustes distance while widely separated units represent larger Procrustes distance (Table 2.2). Overall, the morphological distances between the Traverse Group formations concur with the stratigraphic succession of the Traverse Group (Fig. 2.5).

Procrustes pairwise distances were plotted against time calculated from stratigraphic thicknesses of Traverse Group strata (Tables 2.3, 2.4, Fig. 2.6).

Maximum-likelihood estimation of the rate and mode of evolution of *P. cf. lineata* dorsal valve shape given the Traverse group stratigraphy, yielded a rate of $0.36 \pm 0.16$ Procrustes units (the standard measure of geometric shape change) per million years and a mode coefficient $a$ of $0.22 \pm 0.24$, indicating near stasis to random mode of shape evolution for the means of this atrypid species lineage. *P. cf. lineata* ventral valve shape yielded a rate of $0.15 \pm 0.10$ Procrustes units per million years and a mode coefficient $a$ of $0.37 \pm 0.12$ indicating near random divergence mode of shape evolution. *P. cf. lineata* anterior of the shell shape yielded a rate of $0.31 \pm 0.09$ Procrustes units per million years and a mode coefficient $a$ of $0.01 \pm 0.14$ indicating stasis. *P. cf. lineata* posterior of the shell shape yielded a rate of $0.50 \pm 0.21$ Procrustes units per million years and a mode coefficient $a$ of $0.11 \pm 0.31$ indicating near stasis to random mode of shape evolution (Fig. 2.6).

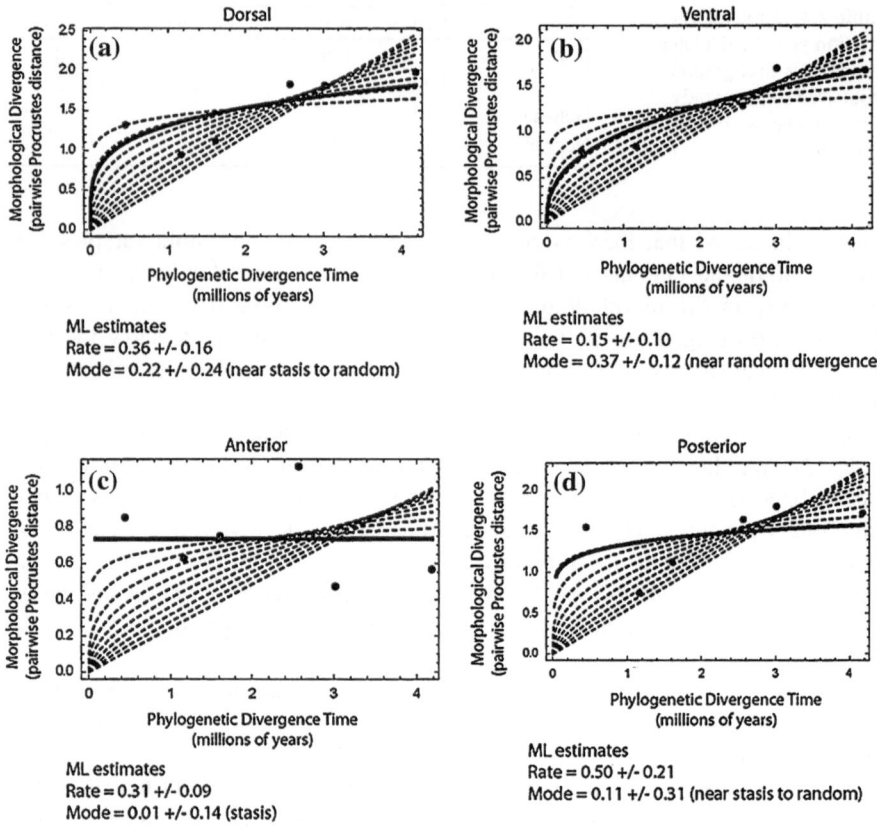

**Fig. 2.6** Graph showing morphometric divergence (pairwise Procrustes distances) and phyloge-netic divergence time (millions of years). The series of *dashed lines* show the expected relation-ship between morphological and phylogenetic divergence time from strong stabilizing selection (0.1), through random divergence (0.5), to diversifying (directional) selection (1.0). The maxi-mum-likelihood estimate of this relationship, shown by the *dark line*, suggests that *P. cf. lineata* has experienced **a** near stabilizing selection to random divergence in dorsal valves, **b** near random divergence in ventral valves, **c** stabilizing selection in anterior margin area of valves and **d** near stabilizing selection to random divergence in posterior hinge area (modified from Bose 2012)

Mean principal component scores of the *P. cf. lineata* dorsal valves regressed on water depth values show a high positive correlation with insignificant results ($r = +0.79$, $p > 0.01$); for those of ventral valves show a very high positive correla-tion with significant results ($r = +0.99$, $p < 0.01$); for those of anterior margin area show a moderate positive correlation with insignificant results ($r = +0.53$, $p > 0.01$) and posterior hinge area shows a low positive correlation with insignificant results ($r = +0.22$, $p > 0.01$). Thus, overall, only the ventral valve morphological shape of this species shows a strong correspondence with the ranges of water depth temporally (Table 2.5). While other morphological shape data (dorsal valve, anterior, and poste-rior areas) show a low to moderately high linear correlation with water depth ranges, they do not show any significant correspondence with changing water depth.

**Table 2.5** Multivariate analysis of variance between morphological principal component scores of individual samples corresponding to medium and shallow water depths

|  |  | Mean water depth (m) | Correlation | P values |
|---|---|---|---|---|
| | *PC1 (dorsal)* | | | |
| Norway Point | −0.99918424 | 1.95 | r = −0.14 | $p = 1.156 \times 10^{-96}$ |
| Genshaw Fm. | 0.457945261 | 31.95 | | MANOVA |
| | | | | $p = 4.74 \times 10^{-07}$ |
| | | | | Regression |
| Ferron Point | −0.1882791 | 39.3 | | r = −0.15 |
| Bell Shale | −0.08658585 | 34.35 | | |
| | *PC1 (ventral)* | | | |
| Norway Point | −0.99427466 | 1.95 | r = 0.3267 | $p = 5.49 \times 10^{-22}$ |
| Genshaw Fm. | 0.09912704 | 31.95 | | MANOVA |
| | | | | $p = 2.57 \times 10^{-29}$ |
| Ferron Point | 0.275272864 | 39.3 | | Regression |
| Bell Shale | 0.118146437 | 34.35 | | |
| | *PC1 (anterior)* | | | |
| Norway Point | 0.561290195 | 1.95 | r = −0.11455 | $p = 4.17 \times 10^{-16}$ |
| Genshaw Fm. | −0.41474042 | 31.95 | | MANOVA |
| Ferron Point | 0.34338073 | 39.3 | | p = 0.0001238 |
| Bell Shale | 0.058519641 | 34.35 | | |
| | *PC1 (posterior)* | | | |
| Norway Point | −0.51941632 | 1.95 | r = −0.07 | $p = 2.65 \times 10^{-75}$ |
| Genshaw Fm. | 0.646570066 | 31.95 | | MANOVA |
| | | | | p = 0.0087 |
| Ferron Point | −0.6960508 | 39.3 | | |
| Bell Shale | −0.11861116 | 34.35 | | |

Statistical correlation (linear regression) between mean morphological principal component (PC axes) scores and water depth values retrieved from Fig. 2.5 of Wylie and Huntoon (2003). 'r' values indicate the linear correlation coefficient between the two variables (mean PC scores of individual samples and water depth). *P* values report the probability that morphological shape change corresponds with the change in water depth

## 2.5 Discussion

### 2.5.1 Qualitative and Quantitative Recognition of Taxa

*Pseudoatrypa* is a common brachiopod from the Givetian to late Frasnian of North America. This genus occurs throughout much of the Traverse Group in the Michigan Basin, including the Bell, Ferron Point, Genshaw, Norway Point, and Potter Farm formations (Kelly and Smith 1947; Koch 1978). Here we focus on material from the Bell Shale, Ferron Point, Genshaw, and Norway Point of the northeastern outcrop of Michigan.

Webster (1921) first described the taxon *Pseudoatrypa* as *Atrypa lineata* from the late Givetian-early Frasnian Cedar Valley Group of Iowa, material which was later recognized as *Pseudoatrypa lineata* species of the new genus

*Pseudoatrypa* (Day 1992; Day and Copper 1998). The holotype of *Atrypa line-ata* (Webster 1921) came from the upper Osage Springs and Idlewild members of the Lithograph City Formation of Iowa (Stumm 1951; Day 1996). Samples in this study from all the four stratigraphic intervals of the Givetian age Traverse Group agree well with the overall morphology of type specimen of *P. lineata* and were called *Pseudoatrypa cf. lineata* for the purpose of this study. These samples are characterized by shell maximum width of 2.1–3.3 ± 0.2 cm, width almost equal to the length of the shell, subquadrate shell shape, dorsibiconvex-convexiplanar shell, flattened with/without umbonal inflation in ventral valves, fine to coarse ribs with implantations and bifurcations, 1–2 plicae/1 mm spacing, somewhat consistent 2–4 mm spacing between growth lines with their crowding at the anterior margin. However, based on morphometric results derived from quantification of morphological shape, the Norway Point samples appear to be different from the lower Traverse Group samples.

When quantitatively examined, the samples of this species from the Traverse group stratigraphic intervals overlap in morphological shape with significant statistical differences between them, in addition to morphological variation within samples from each interval. Principal component analysis indicates morphologies of *P. cf. lineata* samples show considerable overlap, which is also observed in overall phenotypic traits of these specimens (Figs. 2.3, 2.4). Notably, samples show some deviation in morphological shape in the uppermost sampled strata of the Traverse group. However, statistically significant morphometric differences clearly exist between the mean shapes of the samples from these intervals using MANOVA tests ($p < 0.01$) and as illustrated by CVA (Fig. 2.4). Thin Plate Spline Real warp plots show that these differences were in the shape of the shell from the mid-point of the specimen to the anterior commissure on both dorsal and ventral valves (landmark points 3, 4, and 5) suggesting shape deflection in the whole anterior region of the shell as these are bilaterally symmetrical organisms. However, this deflection was not so much, so as to count for change. Further tests show that anterior region of the shells remained static through time unlike dorsal valve and posterior region that were near static to randomly diverging through time. Ventral valve randomly diverged in time. This, further, suggests that anterior margin was the most stable in contrast to posterior hinge, dorsal, and ventral valve area. Anterior margin stability reflects a functional significance as evident from studies related to frills in atrypids where they are known to stabilize the shells in the substrate (Copper 1967).

## 2.5.2  Punctuated Equilibrium Hypothesis

The punctuated equilibrium model of evolution predicts that the morphology of a species will be relatively static during most of the species history of evolution with change occurring at rapid and episodic events of speciation (Eldredge and Gould 1972; Gould and Elredge 1977). In this case study, *Pseudoatrypa cf. lineata* species lineage was traced in the Traverse Group stratigraphic intervals of Michigan to test the punctuated equilibrium model, in fact to test for stasis, since it persisted through that time range. If this model is true, then very little or no morphological change is expected between samples in successive strata as this is only one species lineage with no evidence of species splitting. However, if data could be retrieved with greater stratigraphic resolution, that is, if samples could be collected from numerous successive finer scale units, then the chances are that the morphological trend could be more reasonable evaluated. Morphological shape trend illustrated for four broad scale stratigraphic units along the major principal component axes shows a large degree of morphological overlap between the lower three stratigraphic intervals (Bell Shale, Ferron Point, and Genshaw Formation) with some deflection in Norway Point samples (Fig. 2.3a–c). MANOVA and CVA show significant statistical differences between samples from all four stratigraphic intervals ($p < 0.01$) (Fig. 2.4); however, the morphological variation within samples in individual units also shows considerable overlap. This suggests, that while there is phenotypic variation within *Pseudoatrypa cf. lineata* itself, there is still considerable overlap in their extent of morphological variation among stratigraphic intervals. Incorrect recording of spacing between stratigraphic units due to lack of absolute ages and presence of stratigraphic gaps often gives an impression of abrupt change within a species (Sheldon 1987). In this study, samples were collected from units with known differential spacing between them and with almost rarely present stratigraphic gaps between them (Fig. 2.1, Table 2.4); however, there is a lack of sampling between Genshaw and Norway Point formations which equals 42.0 m (125.0 feet) in thickness. Thus, while the abrupt deflection in morphological variation in Norway Point formation may appear to be due to sampling bias to some extent, it cannot be further proved without illustrating the morphological trend of the samples from the intermediate stratigraphic units between Genshaw and Norway Point formations. This change may have to do with some kind of anagenetic change in morphological shape pattern within this species lineage as proposed in prior studies by Roopnarine and others (1999).

Lieberman et al. 1995 observed morphological overlap between the lowermost and uppermost occurrences with some variations in the intervening samples within *Mediospirifer audaculus* and *Athyris spiriferoides* species lineages sampled from successive stratigraphic horizons of the Hamilton Group (5 million year) of New York. They performed this study using traditional morphometrics and interpreted the results as a case of stasis without performing real tests for determining evolutionary modes within those two species lineages. In this study, morphological distances retrieved from geometric morphometric analysis were further used to plot morphological shape against evolutionary divergence time. *P. cf. lineata* samples, an atrypid species lineage was studied using geometric morphometric technique, from a 6 million year time interval, from the Middle Devonian period that has never been tested for stasis or other evolutionary patterns. Thus, whether results testing evolutionary patterns support the stasis hypothesis or not, is discussed under another section, called rate and mode of evolution.

## 2.5.3 Phyletic Gradualism Hypothesis

The phyletic gradualism hypothesis is that species continue to adapt to new environmental and biological selection pressures over the course of their life history, gradually becoming new species. During this process, chances are that anagenetic evolution occurs at a smooth, steady, and incremental (not necessarily constant or slow) rate on a geological timescale (Darwin 1859). Sheldon (1987) reported evidence of parallel gradualistic evolution in eight lineages of 15,000 Ordovician trilobites from central Wales over a 3 million year interval based on evidence of increasing number of pygidial ribs in more than one species lineage. However, in a later study, Sheldon (1996) proposed the Plus ~ca change model that predicts a tendency for continuous, gradualistic evolution in narrowly fluctuating, relatively stable environments, and more stasis with episodic punctuations in widely fluctuating environments with its application in more physical environmental variables like sea level, substrate, temperature, etc., on a geologic time scale (Sheldon 1990, 1996). Nevertheless, the traditional conception of Darwin's gradualistic evolution has been persisting as the mainstream idea among evolutionary biologists through the last two centuries. However, observed patterns in the fossil record alone are not sufficient to confirm the theory of phyletic gradualism; the mechanisms behind the pattern must also be described and demonstrated (Gould and Eldredge 1977). Therefore, while some fossil evidence supports Darwin's more traditional concept of phyletic gradualism, overwhelming evidence from the fossil record and theoretical studies now implies that punctuated equilibrium (punctuation and stasis) is the foremost pattern in macroevolution (Eldredge and Gould 1972; Gould and Eldredge 1977; Williamson 1981; Stanley and Yang 1987; Gould 1988; Barnosky 1990; Lich 1990). Here we tested for evolutionary patterns in a single species lineage, to see if stasis or change predominates the Traverse Group stratigraphic units.

In this study, close morphological links were established in *P. cf. lineata* taxon among the three lowermost successive strata (Bell Shale, Ferron Point, and Genshaw Formation) with some distance with the uppermost strata (Norway Point) under investigation. The morphometric shape distances concur with the stratigraphic arrangement of the Traverse Group (Fig. 2.5). Results from Euclidean cluster analysis in part appear to support the phyletic gradualism hypothesis in that the closely spaced intervals show smaller morphological distances than those widely spaced in time. However, morphological trend alone cannot give evidence for gradual evolution in this species over time. While morphological shape change appears to be oscillating around the mean and morphological distances appear to be incremental in time, a more strong evidence is required that tests the rate and mode of evolution to prove that this shape change pattern is driven by directional selection.

## 2.5.4 Tempo and Mode of Evolution

The evolution of dorsal valve, ventral valve, anterior margin, and posterior hinge shape morphology in *Pseudoatrypa cf. lineata* is investigated under three different evolutionary modes: (1) stabilizing selection; (2) randomly fluctuating selection; and (3) directional selection. Stabilizing selection, is a type of selection, in which the morphology in the given unit is adjusted or pushed back with respect to the earlier morphologies, in cases where it moves too far away from the mean morphology. Randomly fluctuating selection is a type of random walk, or Brownian motion, in which the direction and magnitude of change in any given generation or time is not influenced by that in earlier or later ones. Directional selection, is the type of selection, in which selection often causes changes in one direction than another (Polly 2004).

The results of the analysis of evolutionary rate and mode in this case study indicate that stabilizing to near random divergence has probably been acting on these *P. cf. lineata* taxon, which concurs with the very small morphological divergences among them (Fig. 2.6). The large degree of morphometric overlap among the *P. cf. lineata* individuals from the four stratigraphic units of Traverse Group might be interpreted by some to represent stasis, since mean shell shape among none of the strata have unambiguously diverged from each other, except for the Norway Point samples. However, the statistical definition of stasis, or stabilizing selection, is that less morphological divergence has occurred than expected under a random-walk (Brownian motion) model of evolution given the amount of time since divergence. The statistical definition of random divergence is that similar morphological divergence has occurred as expected under a random-walk model of evolution given the amount of time since divergence. For this atrypid taxon, the changes in mean shape tested in four different views are less than or equal to that expected given time since divergence (since lowest strata in Traverse Group succession)—the most likely estimate of $a$ in Eq. 2.1 given the data presented in Fig. 2.6 ranges between <0.01 and $\leq$0.50 (dark line). Stasis would produce a pattern where the best fit would have a value near 0.0 for parameter $a$, and random selection would produce a pattern where the best fit would have a value near 0.5 for parameter $a$. The rate of divergence in this brachiopod species over stratigraphic intervals is not linear as expected with directional selection. The best interpretation for these data is that they produce a pattern which ranges from near stasis to near random mode of evolution.

Stabilizing selection is one of the several processes thought to explain patterns of morphological stasis (Vrba and Eldredge 1984; Smith et al. 1985; Lieberman et al. 1995; Polly 2004). However, later investigation has shown that stasis like patterns is produced when different selection pressures act on species belonging to different ecosystems, overall, producing no net morphological trend (Lieberman et al. 1994, 1995, Lieberman and Dudgeon 1996). When morphological fitness is influenced by many independent environmental variables (e.g., nutrition, winter temperature, and predator density), morphologies can oscillate in time with changing environments (Polly 2004). Thus, randomly fluctuating selection can explain patterns of morphological change in species over time. Here, we have selected samples from more stable environmental regimes with similar lithologic settings, water depth, and sea-level cycle to detect morphological patterns across time. In this study, an evaluation of real morphological distances on brachiopod morphological shape against geologic time suggests that they have evolved predominantly by stabilizing to randomly fluctuating selection. The dorsal valve had chances of evolving both via near stabilizing to random selection, ventral valve via near random selection, anterior margin via stabilizing selection, and posterior hinge via near stabilizing to random selection. This suggests that morphologies from ancestral to descendant populations in this species lineage evolved statically to randomly, which does not support completely either of the evolutionary models (gradual evolution and punctuated equilibrium).

Over 250 documented cases of phenotypic traits evolving within fossil lineages were depicted by Hunt (2007) in his study where he observed only 5 % cases indicated directional evolution, 95 % cases involved random walk and stasis patterns with equal chances for each. In this study, the best interpretation of shape evolution in *P. cf. lineata* taxon, given the data, is that these were statically to randomly evolving over time at a rate slow to moderate enough that they still overlapped considerably through the time period covered in our study. However, morphological shape of descendant populations in Norway Point samples depicts an enormous deflection from the ancestral populations suggesting somewhat anagenetic shape evolution may have occurred in this species later in time. Overall, our morphometric data for the *P. cf. lineata* species lineage are consistent with the stratigraphic succession of Traverse Group.

## 2.5.5  Morphology and Environment

The paleontological record of the lower and middle Paleozoic Appalachian foreland basin demonstrates ecological and morphological stability on geological time scales. Some 70–80 % of fossil morphospecies within assemblages persisted in similar relative abundances in coordinated packages lasting as long as 7 million years despite evidence for environmental change and biotic disturbances (Morris et al. 1995). This phenomenal evolutionary stability despite environmental fluctuations has been explained by the concept of ecological locking. Ecological locking provides the source of evolutionary stability that is suggested to have been caused by ecological interactions that maintain a static adaptive landscape and prevent both the long-term establishment of exotic invading species and evolutionary change of native species species (Morris 1995; Morris et al. 1995). For example, competition plays an important role in mediating stasis by stabilizing selection (Lieberman and Dudgeon 1996).

When disturbance exceeds the capacity of the ecosystem, evolution proceeds at high rates of directional selection during the organization of a new stable ecological hierarchy (Morris et al. 1995). Thus, ecological evolutionary units (EEUs) and subunits (EESUs) have been used to explain why many fossil species appear to persist unchanged morphologically for long periods, punctuated by short bursts of rapid change. My samples are from the Traverse Group of the Michigan Basin that falls within the Hamilton ecological evolutionary subunit, thus, a good case study to determine morphological patterns in a species lineage from an EESU.

A better understanding is required as of how environment can affect the morphology of a species. Vrba (1980) found that the ultimate driving force for evolutionary change was the environment that includes both the physical and biotic conditions surrounding an organism. Webber and Hunda (2007) quantified environment from faunal abundance data in each bed of Kope and lower Fairview Formations spanning over 2 million year interval using DCA ordination scores and used it as a proxy for water depth. They observed morphological shape in Upper Ordovician trilobite *Flexicalymene granulosa* changes with change in paleoenvironmental conditions. In contrast, for lineages in a single locality, when environments were more narrowly fluctuating and relatively stable, net directional change and/or more widely fluctuating morphologies were predicted (Sheldon 1996, Fig. 2.2). In this study, *Pseudoatrypa cf. lineata* sample collection was from six different localities in Michigan, confined to the shale units of Bell Shale, Ferron Point, Genshaw and Norway Point formations of the Traverse Group (Fig. 2.1). Water depths in the Michigan Basin during transgressive–regressive cycle If, decreased from approximately 50 to <2 m (Bell Shale through Norway Point deposition) and the eustatic curve, in contrast, depicts overall sea level rise punctuated only by regression during the final subcycle of transgressive–regressive cycle If (Wylie and Huntoon 2003). Shallow water depth and onset of regression in the transgressive–regressive cycle If in the upper Traverse Group, somewhat coincides with the morphological shape fluctuation in the Norway Point samples with greater water depth and overall transgression coinciding with the lower Traverse Group strata. However, the overall range of water depth, 2–50 m, and the Middle Devonian eustatic curve represented by transgression–regression cycle during this time, was considered to be stable in general.

Morphological shape in four different views, when regressed onto the shallowing trend of water depth from basal Bell Shale to upper Norway Point formation, shows a low to high positive correlation with significant results only for ventral valve ($p < 0.01$). Other views do not show significant correlation between the two ($p > 0.01$). While it is challenging to discuss morphological shape change patterns in terms of just one or two environmental variables, it is still quite noteworthy, to see the correspondence of Norway Point samples with shallow water depth (<2 m) and Bell Shale, Ferron Point, and Genshaw Formation samples with relatively deeper (40–50 m) water levels. In other words, the small morphological deflection of Norway Point samples from the lower stratigraphic intervals could be in conjunction with the sudden outburst of transgressive–regressive cycle in the whole transgression cycle If and abrupt shallowing of the water level from 50 to <2 m during deposition of the Norway Point Formation.

Atrypids are phenotypically plastic (Fenton and Fenton 1935), and thus it is very challenging to distinguish between true evolution in a species and phenotypic plasticity. Controlling for environment while sampling, to some extent, may help in interpreting true morphological shifts and evolutionary patterns in this species lineage. In anyway, if change in morphological shape in *P. cf. lineata* was only an ecophenotypic phenomenon, one would expect far closer correlation between morphological shape and water depth variable which is clearly not the case in this study.

## 2.6  Conclusions

Landmark measurements in atrypid species lineage *Pseudoatrypa cf. lineata* from the Middle Devonian Traverse Group of Michigan State have been analyzed in this study to determine whether morphological shape trends in lineage can be explained by phyletic gradualism, punctuated equilibria, and/or ecophenotypic variations. Geometric morphometric and multivariate statistical analyses reveal significant statistical differences in morphological shape between Traverse group stratigraphic units, but a considerable overlap is noticed among morphologies over 6 million year interval of time. Though, the samples from the uppermost strata, Norway Point formation shows an abrupt morphological shift from the lower stratigraphic units, Bell Shale, Ferron Point, and Genshaw Formation. Thus, over a period of 6 million years, morphological shape, a species diagnostic character, underwent very little change in the lower Traverse Group formations with some change been reflected in the upper Norway Point formation.

Maximum-likelihood method suggests slow to moderate rate of evolution with near stasis to random divergence mode of evolution in *P. cf. lineata* species lineage. Overall, the mean shape morphological trend suggests considerable morphological overlap between the successive stratigraphic units of the lower Traverse Group with small morphological oscillations in the species life history. However, samples in the uppermost strata deviate from the mean so far, such that some kind of anagenetic evolution appears to be acting on the descendant populations in this species.

Morphological shape in dorsal valves and posterior hinge area suggests evolution by near stasis to random divergence with similar magnitudes of rates (moderate) of evolution. Morphological shape in ventral valves suggests evolution by random divergence with relatively slower rate of evolution. Morphological shape in anterior margin suggests evolutionary stasis with moderate rate of evolution. Overall, this suggests, that ventral valves evolve at a lesser magnitude than all other shape measurements that include dorsal valves, posterior hinge area, and anterior margin. While ventral valves show maximum fluctuation in their evolutionary mode, anterior margin is most stable in morphology over time.

Changes in morphological shape patterns in *P. cf. lineata* weakly correlates with variation in water depth, except for ventral valves. Thus, this suggests, that the morphological shape trend can be explained by stabilizing selection and/or by randomly fluctuating selection, and not by ecophenotypic variations. Since other environmental variables were not tested, it was challenging to infer if any abiotic or biotic factors were behind these mechanisms of evolutionary selection.

Thus, the results from this study, in a complete sense, are neither in full favor of the punctuated equilibrium model nor the phyletic gradualism model. Static evolution is represented in the species early life history with anagenetic evolution, most likely, predominating the later stages in the species life.

# References

Barnosky AD (1990) Evolution of dental traits since latest Pleistocene in meadow voles from Virginia. Paleobiology 16:370–383

Bookstein FL (1989) Principal warps: thin-plate splines and the decomposition of deformations. IEEE Trans Pattern Anal Mach Intell 11:567–585

Bookstein FL (1991) Morphometric tools for landmark data: geometry and biology. Cambridge University Press, Cambridge 435 p

Bose R (2012) Biodiversity and evolutionary ecology of extinct organisms. Paleoenvironmental Sciences. Springer, New York 100 p

Brett CE (1986) The Middle Devonian Hamilton group of New York: an overview. In: Brett C (ed) Dynamic stratigraphy and depositional environments of the Hamilton Group (Middle Devonian) in New York State, Part 1. New York State Museum Bulletin, vol 457, pp 1–4

Brett CE, Baird GC (1995). Coordinated Stasis and Evolutionary Ecology. In: Erwin DH, Anstey RL (eds) New approaches to Speciation. Columbia University Press, New York, pp 285–315

Cooper GA, Butts C, Caster KE, Chadwick GH, Goldring W, Kindle EM, Kirk E, Merriam CW, Swartz FM, Warren PS, Warthin AS, Willard B (1942) Correlation of the Devonian sedimentary formations of North America. Bull Geol Soc Am 53:1729–1794

Copper P (1967) Adaptations and life habits of Devonian atrypid brachiopods. Palaeogeogr Palaeoclimatol Palaeoecol 3:363–379

Darwin C (1859) On the origin of species by means of natural selection, or the preservation of favoured races in the struggle for life. John Murray, London

Day J (1992) Middle-Upper Devonian (late Givetian-early Frasnian) brachiopod sequence in the Cedar Valley Group of central and eastern Iowa. In: Day J, Bunker BJ (eds) The stratigraphy, paleontology, depositional and diagenetic history of the Middle-Upper Devonian Cedar valley group of central and eastern Iowa, vol 16. Iowa Department of Natural Resources Guidebook Series, pp 53–105

Day J (1996). Faunal signatures of Middle-Upper Devonian depositional sequences and sea level fluctuations in the Iowa Basin: U.S. midcontinent. In: Witzke BJ, Ludvigson GA, Day J (eds) Paleozoic sequence stratigraphy, views from the North American Craton. Geological Society of America Special Paper, vol 306, pp 277–300

Day J, Copper P (1998) Revision of latest Givetian-Frasnian Atrypida (Brachiopoda) from central North America. Acta Palaeontol Pol 43:155–204

Ehlers GM, Kesling RV (1970) Devonian strata of Alpena and Presque Isle counties. Michigan Basin Geological Society, Michigan, p 130

Eldredge N, Gould SJ (1972) Punctuated equilibria. In: Schopf (ed) Models in Paleobiology. Freeman, Cooper, pp 82–115

Ettensohn FR (1985) The Catskill Delta complex and the Acadian orogeny: a model. In: Woodrow DL, Sevon WD (eds) The Catskill Delta. Geological Society of America Special Paper 201. Boulder, Colorado, pp 39–50

Felsenstein J (1988) Phylogenies and quantitative characters. Annu Rev Ecol Syst 19:445–471

Fenton CL, Fenton MA (1935) Atrypae described by Clement L. Webster and related forms (Devonian, Iowa). J Paleontol 9:369–384

Geary DH (1995) Investigating species-level transitions in the fossil record: the importance of geologically gradual change. In: Erwin DH, Anstey RA (eds) New approaches to speciation in the fossil record. Columbia University Press, New York, pp 67–86

Geary DH, Hunt G, Magyar I, Schreiber H (2010) The paradox of gradualism: phyletic evolution in two lineages of lymnocardiid bivalves (Lake Pannon, central Europe). Paleobiology 36:592–614

Gingerich PD (1993) Quantification and comparison of evolutionary rates. Am J Sci 293A:453–478

Gingerich PD (1976) Paleontology and phylogeny: patterns of evolution at the species level in early Tertiary mammals. Am J Sci 276:1–28

Gingerich PD (1985) Species in the fossil record: concepts, trends and transitions. Paleobiology 11:27–41

Gingerich PD (2001) Rates of evolution on the time scale of the evolutionary process. Genetica 112–113:127–144

Goldman D, Mitchell CE (1990) Morphology, systematics, and evolution of Middle Devonian Ambocoeliidae (Brachiopoda), Western New York. J Paleontol 64:79–99

Gould SJ, Eldredge N (1977) Punctuated Equilibria: the tempo and mode of evolution reconsidered. Paleobiology 3:115–151

Gould SJ (1988) Prolonged stability in local populations of *Cerion agassizi* on Great Bahama bank. Paleobiology 14:1–18

Hunt G (2007) The relative importance of directional change, random walks, and stasis in the evolution of fossil lineages. Proc Natl Acad Sci 104:18404–18408

Isaacson PE, Perry DG (1977) Biogeography and morphological conservatism of *Tropidoleptus* (Brachiopoda, Orthida) during the Devonian. J Paleontol 51:1108–1122

Kelly WA, Smith GW (1947) Stratigraphy and structure of Traverse Group in Afton-Onaway area, Michigan. Bull Am Assoc Petrol Geol 31:447–469

Kesling RV, Segall RT, Sorensen HO (1974) Devonian strata of Emmet and Charlevoix counties, Michigan. Michigan Museum of Paleontology Papers on Paleontology, vol 7, pp 1–187

Koch WF (1978) Brachiopod paleoecology, paleobiogeography, and biostratigraphy in the upper middle Devonian of Eastern North America: an ecofacies model for the Appalachian, Michigan, and Illinois basins. Dissertation, Oregon State University, Oregon

Lande R (1976) Natural selection and random genetic drift in phenotypic evolution. Evolution 30:314–334

Lich D (1990) *Cosomys primus*: a case for stasis. Paleobiology 16:384–395

Lieberman BS, Brett CE, Eldredge N (1994) Patterns and processes of stasis in two species lineages from the Middle Devonian of New York State. American Museum of Natural History Novitates, vol 3114, New York, NY

Lieberman BS, Brett CE, Elredge N (1995) A study of stasis in two species lineages from the Middle Devonian of New York State. Paleobiology 21:15–27

Lieberman BS, Dudgeon S (1996) An evaluation of stabilizing selection as a mechanism for stasis. Palaeogeogr Palaeoclim Palaeoecol 127:229–238

Mayr E (1963) Animal species and evolution. Harvard University Press, Cambridge

Macleod N, Forey PL (2002) Morphology, shape, and phylogeny. Taylor and Francis, New York

Morris PJ (1995) Coordinated stasis and ecological locking. Palaios 10:101–102

Morris PJ, Ivany LC, Schopf KM, Brett CE (1995) The challenge of paleoecological stasis: reassessing sources of evolutionary stability. Proc Natl Acad Sci 92:11269–11273

Polly PD (2001) Paleontology and the comparative method: ancestral node reconstructions versus observed node values. Am Nat 157:596–609

Polly PD (2004). On the simulation of morphological shape: mutivariate shape under selection and drift. Palaeontol Electron 7:L7A (Coquina Press)

Polly PD (2008) Adaptive zones and the pinniped ankle: a three-dimensional quantitative analysis of carnivoran tarsal evolution. In: Sargis E, Dagosto M (eds) Mammalian evolutionary morphology: a tribute to Frederick S. Szalay. Springer, Dordrecht, pp 167–196

Rohlf FJ (1990) Fitting curves to outlines. In: Proceedings of the Michigan morphometrics workshop, pp 167–177

Rohlf FJ, Slice DE (1990) Extensions of the procrustes method for the optimal superimposition of landmarks. Syst Zool 39:40–59

Rohlf FJ (1999) Shape statistics: procrustes superimpositions and tangent spaces. J Classif 16:197–223

Rohlf FJ, Marcus LF (1993) A revolution in morphometrics. Trends Ecol Evol 8:129–132

Roopnarine PD, Byars G, Fitzgerald P (1999) Anagenetic evolution, stratophenetic patterns, and random walk models. Paleobiology 25:41–57

Roopnarine PD (2001) The description and classification of evolutionary mode: a computational approach. Paleobiology 27:446–465

Roopnarine PD (2003) Analysis of rates of morphologic evolution. Annu Rev Ecol Evol Syst 34:605–632

Sheldon PR (1987) Parallel gradualistic evolution of Ordovician trilobites. Nature 330:561–563

Sheldon PR (1990) Shaking up evolutionary patterns. Nature 345:772

Sheldon PR (1996) Plus qa change—a model for stasis and evolution in different environments. Palaeogeogr Palaeoclimatol Palaeoecol 127(209):227

Simpson GG (1953) The major features of evolution. Columbia University Press, New York

Slice DE (2001) Landmark coordinates aligned by procrustes analysis do not lie in Kendall's shape space. Syst Biol 50:141–149

Smith MJ, Burian R, Kauffman S, Alberch P, Campbell J, Goodwin B, Lande R, Raup D, Wolpert L (1985) Developmental constraints and evolution. Quart Rev Biol 60:265–287

Stanley SM (1979) Macroevolution. WH Freeman, San Francisco

Stanley SM, Yang X (1987) Approximate evolutionary stasis for bivalve morphology over millions of years: a multivariate, multilineage study. Paleobiology 13:113–139

Stumm EC (1951) Check list of fossil invertebrates described from the middle Devonian traverse group of Michigan. Contrib Mus Paleontol Univ Mich 9:1–44

Vrba ES (1980) Evolution, species and fossils: how does life evolve? S Afr J Sci 76:61–84

Vrba ES, Eldredge N (1984) Individuals, hierarchies and processes: towards a more complete evolutionary theory. Paleobiol 10:146–171

Webber AJ, Hunda BR (2007) Quantitatively comparing morphological trends to environment in the fossil record (Cincinnatian series; Upper Ordovician). Evolution 61:1455–1465

Webster CL (1921) Notes on the genus *Atrypa*, with description of new species. Amer Midl Nat 7:13–26

Williamson PG (1981) Paleontological documentation of speciation in Cenozoic mollusks from Turkana Basin. Nature 293:437–443

Wylie AS, Huntoon JE (2003) Log-curve amplitude slicing: Visualization of log data and depositional trends in the Middle Devonian Traverse Group, Michigan basin, United States. Am Assoc Pet Geol Bull 87:581–608

Zelditch ML, Swiderski DL, Sheets HD, Fink WL (2004) Geometric morphometrics for biologists, A primer. Elsevier Academic Press, New York

# Chapter 3
# Comparative Trends in Evolution Across Correlated Geological Strata

**Abstract** While many Devonian correlated stratigraphic units are yet to be tested for species evolution, the middle Devonian Hamilton Group and Traverse Group have been studied to investigate the same. Fossil brachiopods were used as tools that allowed close investigation of the evolutionary response of these species across correlated strata in Eastern America. Results show that their evolutionary patterns are uncoordinated in nature. Thus, this study reveals that evolutionary patterns in brachiopod species lineages do not seem to match in correlated strata at the regional scale in the United States. Abrupt change was noted in the upper strata of the Traverse Group of Michigan samples from the lower ones. This could be interpreted as the origin of a new species, either from environmental selection pressure or by an immigration event. Comparison of Michigan Basin sections with the contemporary Appalachian Basin sections suggests that evolutionary patterns are not linked in nature. Morphologies from the uppermost units in the Traverse Group show sudden deviation from the lowermost units, unlike Hamilton Group where morphological overlap was prominent between the lowermost and uppermost units. Thus, morphological trend in a fossil brachiopod species lineage in the Michigan Basin appears to be local in scope. Other correlated Devonian strata like the Silica Shale of Ohio and the Arkona Shale of Ontario are yet to be tested for species evolution.

## 3.1 Introduction

Correlation studies have gained significant importance in sedimentology, stratigraphy and paleontology studies over three decades now (Prell et al. 1986; Sparling 1988; Brett 1995; Veizer et al. 1997; Ellwood et al. 1999; Bartholomew and Brett 2007; Brett et al. 2008; Brett et al. 2010, 2011). Givetian age rocks have been known to preserve the best brachiopod fossils in the Paleozoic era (Boucot et al. 2008; Isaacson 2008; Bose et al. 2011; Bose 2012a, b). Thus, these Givetian stratigraphic successions have laid the platform for testing the evolutionary theories in some extinct brachiopod fossil lineages. In this study, we focus on the middle Devonian Traverse Group and Hamilton Group successions. The Traverse Group

R. Bose and A. J. Bartholomew, *Macroevolution in Deep Time*, SpringerBriefs in Evolutionary Biology, DOI: 10.1007/978-1-4614-6476-1_3, © The Author(s) 2013

**Fig. 3.1** Brachiopod
specimens from Genshaw
Formation of Traverse Group
of Michigan

**Fig. 3.1** Brachiopod specimens from Genshaw Formation of Traverse Group of Michigan

is located on the northeastern flanks of the Michigan Basin, and Hamilton Group is located on the northwestern flanks of the Appalachian Basin. A topographic high called the Findlay Arch separates the Michigan Basin from the Appalachian Basin (Birchard and Risk 1990).

Based on biostratigraphic evidence and sequence stratigraphic analysis, the Traverse ecological evolutionary (EE) subunit correlates with the Hamilton EE subunit (Goldman and Mitchell 1990; Brett and Baird 1995; Bartholomew and Brett 2007; Brett et al. 2010). The Traverse fauna within the Middle Devonian of the Michigan Basin subprovince displays a high level of faunal and compositional persistence and thus is defined as the Traverse EE subunit (Bartholomew 2006). The small-scale community stability of the Hamilton fauna within the Middle Devonian of the Appalachian Basin subprovince is defined as the Hamilton EE subunit (Brett and Baird 1995). Thus, the morphological trends observed in the Traverse Group can be compared to previously studied patterns in the contemporary Hamilton Group.

In this study, brachiopod fossils from Traverse Group of Michigan Basin (Order: Atrypida) (Fig. 3.1) and Hamilton Group of Appalachian Basin (Order: Athyrida and Spiriferida) have been compared to determine the nature of evolutionary response of these species lineages over time.

## 3.2  Geologic Setting

The study region, formerly part of the eastern Laurentian paleocontinent, includes the northwestern margin of the Appalachian foreland basin and the northeastern rim of the Michigan intracratonic basin. During the late Eifelian to middle Givetian, eastern Laurentia lay south of the equator with the northern Michigan Basin positioned at about 25°S latitude and the northern Appalachian Basin about 30–35°S (Witzke and Heckel 1988). Newly uplifting sediment source areas were created during the second tectophase of the Acadian Orogeny, reflecting oblique collision of the Avalonian terrane with eastern Laurentia (Ettensohn 1987). Lithospheric flexure associated with Acadian tectonic loading also produced a retroarc foreland basin in eastern Laurentia that extended from Alabama into Maritime Canada. At various times, a complementary forebulge formed the western rim of this basin and the foredeep–forebulge couplet may have migrated cratonward through the course of Acadian tectonism (Ettensohn 1987). However, details of this paleogeography that remain incompletely understood are not the objective of our study. For the purposes of this study, we focus primarily on the geologic successions of the Michigan intracratonic basin and the Appalachian foreland basin.

At present, the strata of the Appalachian foreland basin are largely separated from those of the Michigan intracratonic basin by a basement arch system, of which the northeastern segment in Ontario is called the Algonquin Arch and the southwestern segment is called the Findlay Arch (Fig. 3.2). Due to erosion of post-Silurian strata over the crest of the Findlay Arch, Hamilton Group strata in Ohio occur in two principal outcrop regions—the east Findlay Arch region (EFA), near Sandusky, and the west Findlay Arch region (WFA), west of Toledo. Hamilton Group outcrops in southwestern Ontario, are restricted to the Chatham Sag, a small saddle located between the Algonquin and Findlay segments of the basement arch system. The Chatham Sag outcrop area includes well-known exposures in the vicinities of Arkona, Thedford, and Ipperwash. The Findlay-Algonquin Arch and Chatham Sag may have existed intermittently during parts of the Paleozoic Era (Fig. 3.2). However, it should not be assumed that a major arch existed in this position during deposition of the Middle Devonian sediments considered herein. In fact, it is argued that there was no such barrier between the two basins, at least in the position of the modern Findlay Arch during deposition of most of the Hamilton-Traverse Group strata. However, a precursor arch may have developed to the southeast of the present Findlay-Algonquin axis during middle to late Givetian time (Fig. 3.2).

### 3.2.1 Biostratigraphy

Biostratigraphic subdivision of strata of the Michigan Basin, and correlation with strata of the Appalachian and Illinois basins, are poorly understood. Here we focus on the stratigraphy of Michigan Basin and its correlation with Appalachian basin strata. Fossil invertebrate taxa have often been used to refine the overall picture of how strata in the different basins relate. Key biostratigraphic taxa include conodonts, brachiopods, stromatoporoids, trilobites, bivalves, and gastropods. The Michigan Basin contains few, if any, biostratigraphically useful cephalopods for the Middle Devonian; zonally important goniatites are known from the Upper Devonian Squaw Bay Limestone that overlies the Traverse Group.

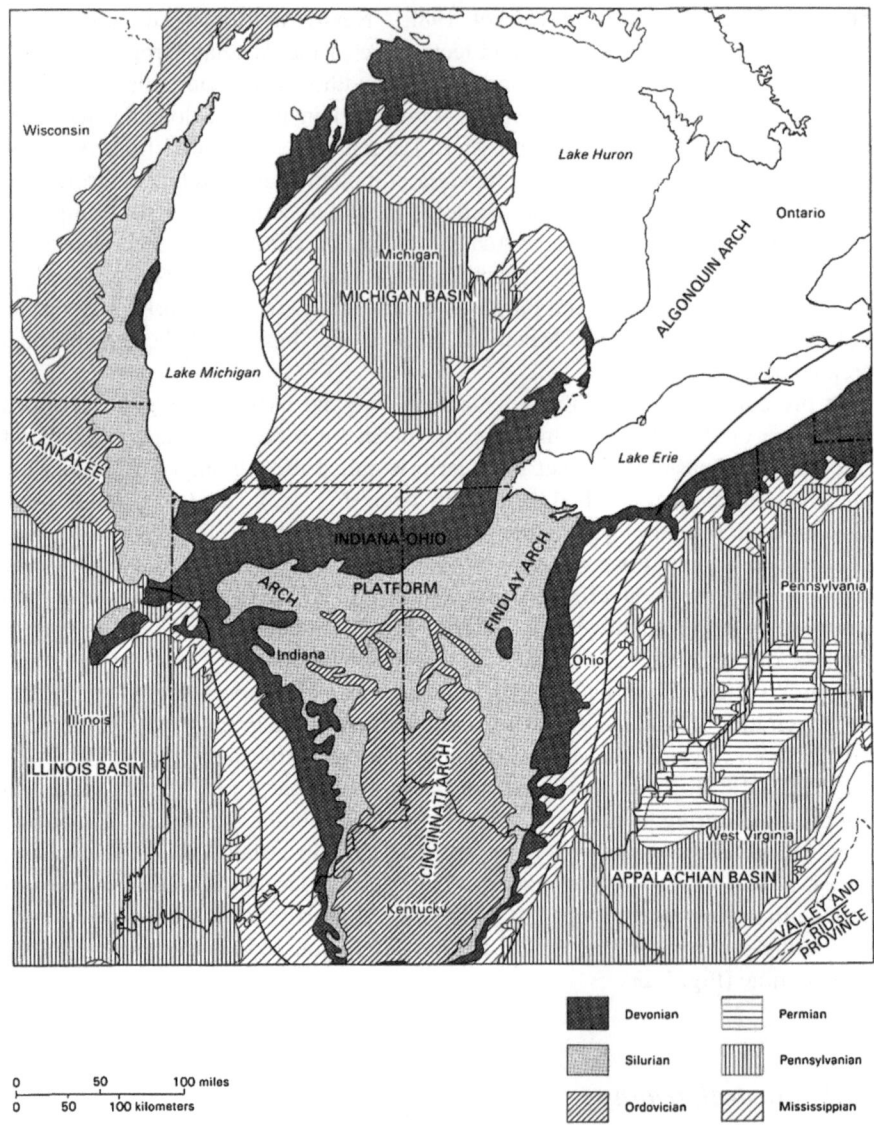

**Fig. 3.2** Regional geologic structures of Michigan Basin and Appalachian Basin (modified from Carlson 1991)

Brachiopod taxa of biostratigraphic importance and of use for correlation in the Middle Devonian of the Michigan Basin include *Variatrpya*, *Subrensselandia*, *Carinatrpya*, *Callipleura*, *Camerophoria*, *Camerospira*, and *Fimbrispirifer*. *Variatrypa*, *Subrensselandia*, and *Carinatrpya* are found only in the Rogers City Limestone that underlies the Traverse Group of Michigan. These taxa are important as they are only known from strata that underlie the Hamilton fauna of the Appalachian Basin, having been recovered from the Stony Hollow/Cherry Valley interval in New York and elsewhere in the Appalachian and Illinois basins (Cooper and Phelan 1966) providing a well-resolved tie at the base of the study interval across a wide area. *Callipleura*, *Camerophoria*, and *Camerospira* all occur in the Newton Creek Limestone and also occur, along with *Fimbrispirifer*, in the Four Mile Dam Formation. These taxa usually occur only in the transgressive carbonates of the third-order sequences of the Appalachian Basin, namely the Centerfield Member of the Ludlowville Formation, and its equivalents, and in the Portland Point Member of the Moscow Formation in New York. Studies of *Pseudoatrypa* remain unknown from the Hamilton Group successions of New York; however, this taxa has been studied in detail from its correlated strata (Traverse Group) in the Michigan Basin (Bose 2012a, b).

### 3.2.2  Traverse Group Stratigraphy

The lowest formational unit of the Traverse Group, the Bell Shale was named by Grabau (1902) for exposures in clay pits north of the village of Bell, Presque Isle County, Michigan, and consists mainly of clay-rich blue-gray shale with intercalated thin shell beds (Fig. 3.3). The Bell Shale is ~23 m (68 ft) thick in the vicinity of Alpena and over 30 m (100 ft) at Presque Isle; as no complete exposure of the Bell exists, all estimates of thickness come from cores (Ehlers and Kesling 1970). The lower portion of the Bell, exposed along the rim of several quarries developed in the Rogers City Limestone, begins with a thin shale (~15 cm) resting directly on top of the Rogers City Limestone. Overlying the basal shale is a ~30 cm coral and brachiopod-rich crinoidal packstone; a large percentage (~50 %) of taxa known from units of the Traverse Group overlying the Bell Shale first appear in this bed, analogous to the Halihan Hill Bed in the Appalachian Basin. The lower portion of the Bell Shale overlying the basal packstone consists of relatively barren shale with thin, bedding-plane accumulations of shell debris composed primarily of bivalve material. The upper portion (~3 m) of the Bell is exposed at the abandoned Kelly's Island Quarry in drainage ditches where it consists of relatively fossil-rich calcareous shales. The Bell Shale is known to thin to the southwest in the subsurface until it disappears in areas to the southwest of a line drawn roughly from Muskegon to Monroe counties (Wylie and Huntoon 2003). To the southeast, the Bell Shale maintains its thickness into the subsurface of southwestern Ontario.

The Ferron Point Formation, named by Warthin and Cooper (1935), consists primarily of blue-gray shale with interbedded fossil-rich pack- to grainstones with a thickness of nearly ~13 m (43 ft) in the Alpena region (Fig. 3.3). As with the Bell Shale, no complete exposures of the Ferron Point Formation exist in Michigan. The lower portion of the Ferron Point is exposed at the Rockport Quarry type section and consists of ~6 m (18 ft) of mainly blue-gray shales with thin, fossil-rich limestones. The upper portion of the Ferron Point Formation, exposed below the basal limestones of the Genshaw Formation at the Ferron Point clay pits off Monaghan Point Road, is almost

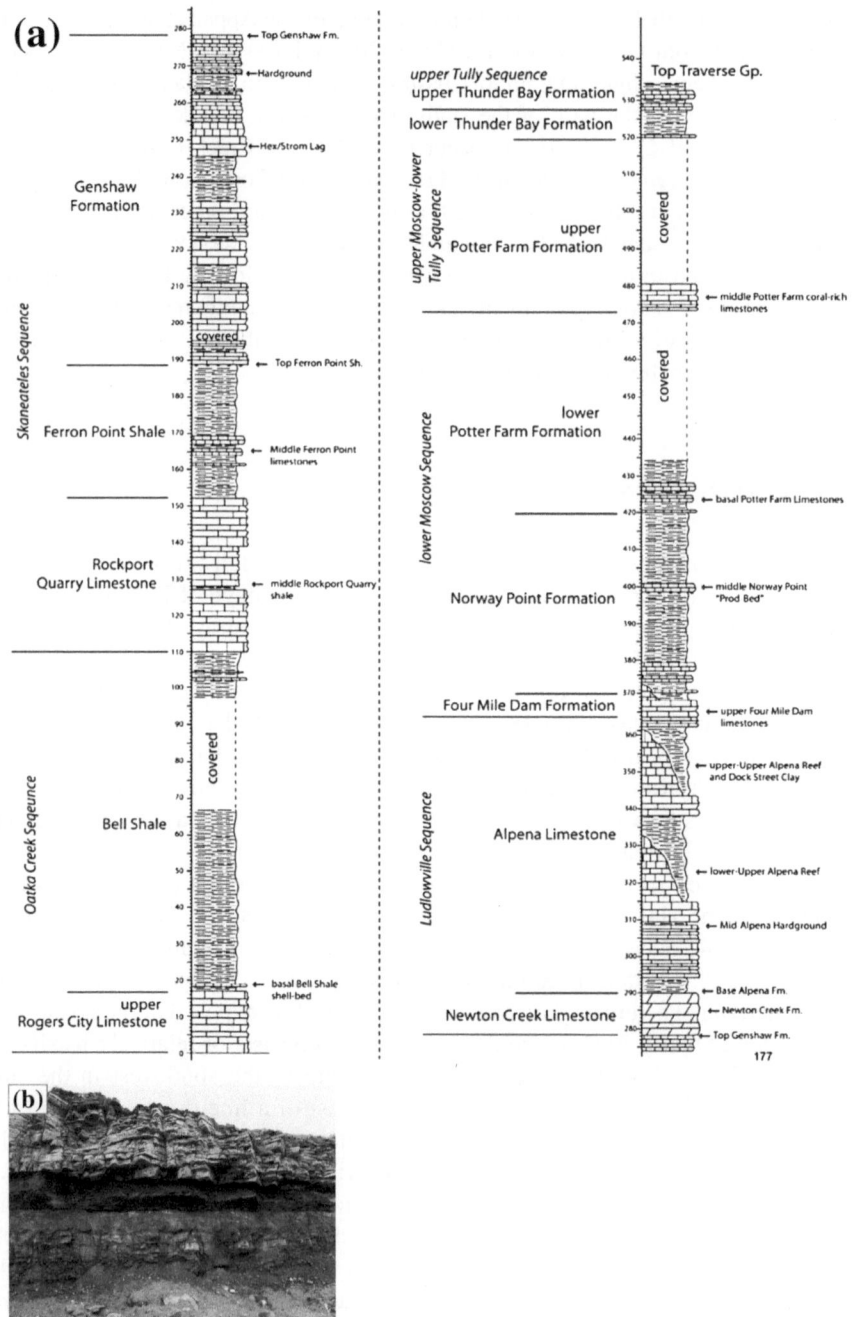

**Fig. 3.3 a** Stratigraphic column for the Middle Devonian Traverse Group in the northeastern Michigan Basin showing formational names and third-order sequence stratigraphic interpretations in italics; **b** field outcrop showing a few stratigraphic beds of the Traverse Group in LaFarge Quarry, Alpena County

wholly composed of greenish blue-gray shale with thin, bedding-plane accumulations of shells. The middle ~8 ft of the Ferron Point is nowhere exposed in the Alpena region. The lowermost Ferron Point Formation is also exposed above the Rockport Quarry Limestone at Black Lake. The unit is known to thin westward down to nearly 8 ft in the Afton-Onoway area (Ehlers and Kesling 1970) and is absent further west.

The Genshaw Formation was named by Warthin and Cooper (1935) for strata lying between their Ferron Point Formation below and their Killians Limestone above, later revision by the same workers (Warthin and Cooper 1943) included the Killians Limestone as the upper member of the Genshaw Formation with the upper contact lying at the base of the overlying Newton Creek Limestone (Fig. 3.3). This formation remained one of the least well-known units of the lower Traverse Group until quite recently when the LaFarge Quarry in the Alpena area began to mine lower into this unit exposing nearly the entire thickness of the formation. The Genshaw Formation can be subdivided into three portions: lower, middle, and upper (formerly the Killians Member); the entire unit has a measured thickness of 30 m (~90 ft at the LaFarge Quarry, Fig. 3.2). West of the Alpena area, the Genshaw Formation records deposition in restricted lagoons, being known only from the subsurface west of the Afton-Onaway area (Wylie and Huntoon 2003). The lower division of the Genshaw Formation begins with a 0.5 m-thick crinoidal grainstone in places displaying burrow prods on its lower surface in contact with the upper Ferron Point Formation. Overlying the basal bed of the Genshaw is a thin shalier succession followed by a limestone-rich interval to the top of the lower Genshaw.

The overlying Norway Point Formation was named by Warthin and Cooper (1935) for strata lying between the Four Mile Dam Formation below and the Potter Farm above. The primary exposures of the unit occur along the Thunder Bay River upstream from Four Mile Dam (Fig. 3.3). The Norway Point is composed of medium-gray, thin-bedded, argillaceous, somewhat fossiliferous limestones in the lower portion of the unit and blue-gray, fossiliferous, clay-rich shales with intercalated siltstone beds in the upper portion of the unit. As only the lower contact of the unit is exposed in a single outcrop where it laps onto the flanks of the bioherm of the underlying Four Mile Dam, the precise thickness of the unit is in question; Ehlers and Kesling (1970) estimated a thickness of about 50–60 ft based on dip and the location of the near-basal exposures of the overlying Potter Farm Formation. The Norway Point Formation is absent west of the Alpena region (Wylie and Huntoon 2003).

Other strata in the Traverse Group—Rogers City Limestone, Rockport Quarry Limestone, Newton Creek Limestone, Alpena Limestone, Four Mile Dam Formation, and Potter Farm Formation remain the subjects of future investigation.

## 3.3  Methods

A comprehensive literature research was performed on the Middle Devonian brachiopod fossils of the Hamilton Group of New York. This search facilitated the comparison of the brachiopod fossil lineages of the Hamilton Group with those of the Traverse Group of Michigan.

The patterns of shape change in these Michigan Basin samples were compared with the patterns observed by previous workers in the Appalachian Basin from the contemporary Hamilton Group units, to which the Traverse units have been correlated based on sequence stratigraphic analysis (Brett et al. 2010).

## 3.4 Results and Discussion

The Michigan Basin is separated from the Appalachian Basin by a basement arch system, the northeastern segment of which is called the Algonquin Arch, and the southwestern segment is called the Findlay Arch (Fig. 3.2). The Traverse Group lies on the flanks of the Michigan Basin and my study area is located in the northeastern part (Northern Peninsula) of the Michigan Basin in Alpena and Presque Isle Counties, Michigan. The Hamilton Group lies on the flanks of the Appalachian Basin, with samples studied in the past for stasis coming from western and central New York, northwestern part of Appalachian Basin.

Conodont-based correlation and sequence stratigraphic analysis of the Middle Devonian strata from the Michigan and Appalachian Basin shows that the Bell Shale is coeval with the upper Marcellus Shale (Oatka Creek Formation), the Ferron Point and Genshaw formations are coeval with the Skaneateles, and the Norway Point Formation with the lower Windom Member of the Moscow Formation (Brett et al. 2009, 2010) (Fig. 3.4). Thus, the Traverse Group and Hamilton Group fauna thrived during similar geologic time periods (Brett et al. 2009, 2010).

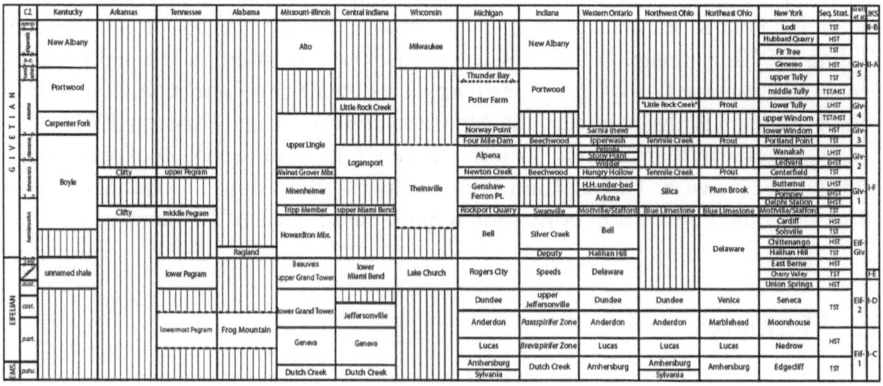

**Fig. 3.4** Conodont-based correlation of Eifelian and Givetian stratigraphic units across Ontario, Ohio, New York, Michigan, Indiana, and other adjacent U.S. states

### 3.4.1 Comparison of the Michigan Basin with the Appalachian Basin

The Traverse Group sections of the Michigan Basin are correlated with the Hamilton Group of western and central New York sections in the Appalachian Basin (Ehlers and Kesling 1970; Brett et al. 2009, 2010) (Fig. 3.5). Patterns of morphological shifts in *Pseudoatrypa cf. lineata* species lineage recorded from the Traverse Group (Chap.2) were compared with the patterns in the *Athyris* and *Mediospirifer* lineages from the Hamilton Group documented by Lieberman et al. (1995). Lieberman et al. (1995) studied samples from various horizons of the Hamilton Group formations (Fig. 3.5). These include Chittenango and Cardiff members of the Upper Oatka Creek formation, Centerfield, Ledyard, Wanakah and Jaycox members of the Ludlowville Formation, and Tichenor-Kashong and Windom members of the Moscow Formation. The upper Oatka Creek Formation is coeval with the Bell Shale, middle-upper Skaneateles Formation with the Genshaw and Ferron Point, and the lower Windom member of the Moscow Formation with the Norway Point of the Traverse Group (Fig. 3.5). Lieberman et al. (1995)

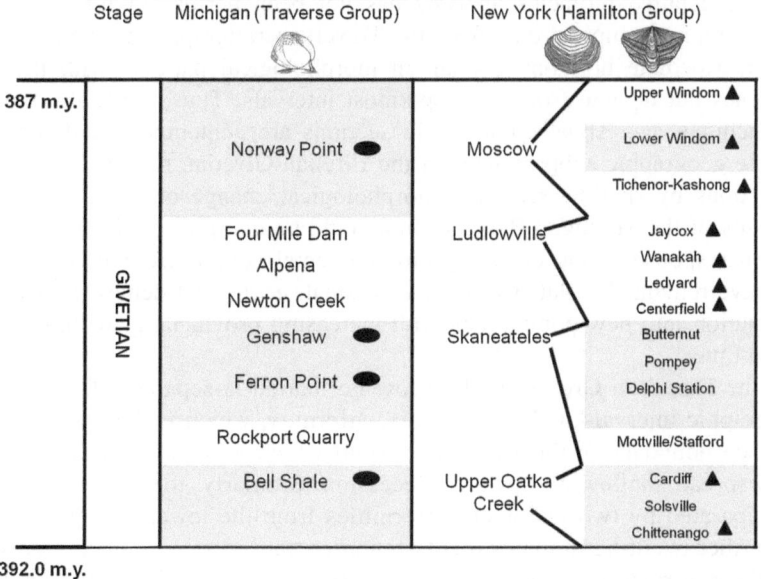

**Fig. 3.5** Chart showing stratigraphic correlation between Traverse Group and Hamilton Group units (modified after Brett et al. 2010). *Shaded region* shows the Traverse Group units sampled and their Hamilton Group equivalents. *Solid triangles* show the sampled horizons from Lieberman et al. (1995) study and *solid circles* show the sampled horizons for this study. *Dark solid lines* for each Hamilton Group formation mark all the members within. A brachiopod sketch of *Pseudoatrypa* sp. is inserted next to the Traverse Group column and brachiopod sketches of *Athyris* sp. and *Mediospirifer* sp. is inserted next to the Hamilton Group column

observed morphological overlap between the lowermost (Oatka Creek Formation) and uppermost (Moscow Formation) occurrences with some variations in the intervening samples.

In this study, atrypid samples show stasis-like patterns in the lower stratigraphic intervals (Bell Shale, Ferron Point, Genshaw) in the Traverse Group with a large change recorded in the uppermost (Norway Point) interval. Samples have not been analyzed from Traverse sections contemporary to the Ludlowville of the Hamilton. Thus, though sampling from intermediate units (Newton Creek and Alpena) of the Traverse may have been ideal for comparing the Michigan and Appalachian Basin sections for accounting morphological patterns in its entirety, the comparative analysis in this study still holds value as the sampled stratigraphic intervals from the Traverse correlates with the lower, upper, and intermediate units of the sampled stratigraphic horizons in the Hamilton Group (Fig. 3.5). The middle stratigraphic horizons of the Hamilton Group has been previously accounted for morphological oscillations with common reversals and considerable overlap in species occurrences in the lowermost and uppermost intervals, while within single biofacies (based on water depth and sedimentation rates), morphological change was evident (Cooper and Dutro 1982; Lieberman et al. 1995).

While detailed biofacies analysis has not been performed for the Traverse Group samples, temporal data from the Traverse stratigraphic horizons show that their intermediate horizons overlap in morphological patterns with the lowermost ones but deviate from the uppermost intervals. Thus, while these units of the Michigan and Appalachian Basin sections are contemporary, though being separate geographic subprovinces in the Eifelian-Givetian time with some interconnections by shallow seas, the morphological change observed in the lower peninsula of the Michigan Basin section could be interpreted as a local phenomenon in response to the changing environmental conditions. Small populations may have remained isolated in their own small ecological demes, thus resulting in evolution into new populations with increasing provinciality within the Basin later in time.

In the Hamilton Group, the Moscow Formation is separated from the lower stratigraphic intervals with a major disconformity, which is the most distinctive sequence boundary in the Hamilton Group which was further terminated with a widespread shallowing-upward succession. Similarly, the Norway Point was also separated by two major unconformities from the lower stratigraphic intervals further resulting in major regression after transgression. Thus, interestingly, both these formations represent the overall shallowest point in the entire succession. Thus, while similar eustatic sea-level changes were noted within the Michigan and Appalachian Basin succession, it is interesting that Lieberman et al.(1995) found considerable morphological overlap between the Norway Point equivalent Moscow Formation and Bell Shale equivalent upper Oatka Creek Formation of the Hamilton succession while this study found morphological dissimilarity between the Norway Point and Bell Shale formations of the Traverse succession.

# 3.5  Conclusions

Comparison of Michigan Basin sections with the contemporary Appalachian Basin sections suggests that morphological trend observed in the brachiopod fossil lineage in the Michigan Basin appears to be local in scope. This is evident from the deflection of the uppermost intervals of the Traverse Group from the lowermost intervals unlike morphological overlap recorded in the Hamilton Group lowermost and uppermost intervals.

# References

Bartholomew AJ (2006) Middle Devonian faunas of the Michigan and Appalachian Basins: comparing patterns of biotic stability and turnover between two paleobiogeographic subprovinces. Dissertation, University of Cincinnati

Bartholomew AJ, Brett CE (2007) Correlation of Middle Devonian Hamilton Group-equivalent strata in east-central North America: implications for eustasy, tectonics and faunal provinciality. Geol Soc Lon Spl Pub 278:105–131

Birchard MC, Risk MJ (1990) Stratigraphy of the Middle Devonian Dundee formation, southwestern Ontario. Ontario Geological Survey Miscellaneous Paper No. 150, pp 71–86

Bose R (2012a) A new morphometric model in distinguishing two closely related extinct brachiopod species. Hist Biol: Int J Paleobiol. doi:10.1080/08912963.2012.658568

Bose R (2012b) Biodiversity and evolutionary ecology of extinct organisms., Earth system sciencesSpringer, New York

Bose R, Schneider CL, Leighton LR, Polly PD (2011) Influence of atrypid morphological shape on Devonian episkeletobiont assemblages from the lower Genshaw formation of the Traverse Group of Michigan: a geometric morphometric approach. Palaeogeogr Palaeoclimatol Palaeoecol 310:427–441

Boucot AJ, Poole FG, Amaya-Martínez R, Harries AG, Sandberg CA, Page WR (2008) Devonian brachiopods of southwesternmost Laurentia: biogeographic affinities and tectonic significance. Geol Soc Am Spl Paper 442:77–97

Brett CE (1995) Sequence stratigraphy, biostratigraphy, and taphonomy in shallow marine environments. Palaios 10(6) (Tenth Anniversary Theme Issue, SEPM—Society for Sedimentary Geology):597–616

Brett CE, Baird GC (1995) Coordinated stasis and evolutionary ecology of Silurian to Middle Devonian faunas in the Appalachian Basin. In: Erwin DH, Anstey RL (eds) New approaches to speciation. Columbia University Press, New York, pp 285–315

Brett CE, Algeo TJ, Mclaughlin PI (2008) Use of event beds and sedimentary cycles in high-resolution stratigraphic correlation of lithologically repetitive successions: the upper Ordovician Kope formation of Northern Kentucky and Southern Ohio. In: Harries PJ (ed) High-resolution approaches in stratigraphic paleontology. Topics in Geobiology, vol 21. Kluwer, Amsterdam, pp 315–350

Brett CE, Ivany LC, Bartholomew AJ, DeSantis MK, Baird GC (2009) Devonian ecological-evolutionary subunits in the Appalachian Basin: a revision and a test of persistence and discreteness. Geol Soc Lond Spl Pub 314:7–36

Brett CE, Baird GC, Bartholomew AJ, DeSantis MK, Straeten CA (2010) Sequence stratigraphy and a revised sea-level curve for the Middle Devonian of eastern North America. Palaeogeogr Palaeoclimatol Palaeoecol 304:21–53

Brett CE, Baird GC, Bartholomew AJ, DeSantis MK, Ver Straeten CA (2011) Sequence stratigraphy and a revised sea-level curve for the Middle Devonian of eastern North America. Palaeogeogr Palaeoclimatol Palaeoecol 304: 21–53

Carlson EH (1991) Minerals of Ohio. Ohio Div Geol Surv Bull 69:1–155

Cooper G, Dutro JT (1982) Devonian brachiopods of New Mexico. Bull Am Paleontol 82–83:1–215

Cooper GA, Phelan T (1966) *Stringocephalus* in the Devonian of Indiana. Smithsonian Misc Collect 151(1):1–21

Ehlers GM, Kesling RV (1970) Devonian strata of Alpena and Presque Isle Counties, Michigan. Michigan Basin. Geological Society Guide Book for Field Trips, 130 p

Ellwood BB, Crick RE, Hassani AE (1999) The magneto-susceptibility event and cyclostratigraphy (MSEC) method used in geological correlation of Devonian rocks from Anti-Atlas Morocco. AAPG Bull 83:1119–1134

Ettensohn FR (1987) Rates of relative plate motion during the Acadian Orogeny based upon the spatial distribution of black shales. J Geol 95:572–582

Goldman D, Mitchell CE (1990) Morphology, systematics, and evolution of Middle Devonian Ambocoeliidae (Brachiopoda), Western New York. J Paleontol 64:79–99

Grabau AW (1902) Michigan. Geological Survey. Annual Report for 1901, p 192

Isaacson PE (2008) Devonian brachiopods from Northeastern Washington: evidence for a non-allochthonous terrane and Late Devonian biogeographic update. Geol Soc Am Spl Paper 442:99–106

Lieberman BS, Brett CE, Eldredge N (1995) A study of stasis in two species lineages from the Middle Devonian of New York State. Paleobiology 21:15–27

Prell WL, Imbrie J, Martinson DG, Morley JJ, Pisias NG, Shackleton NJ, Streeter HF (1986) Graphic correlation of oxygen isotope stratigraphy application to the Late Quaternary. Paleoceanography 1:137–162

Sparling DR (1988) Middle Devonian stratigraphy and conodont biostratigraphy, north-central Ohio. Ohio J Sci 88(1):2–18

Veizer J, Buhl D, Diener A, Ebneth S, Podlaha OG, Bruckschen P, Jasper T, Korte C, Schaaf M, Ala D, Azmy K (1997) Strontium isotope stratigraphy: potential resolution and event correlation. Palaeogeogr Palaeoclimatol Palaeoecol 132:65–77

Warthin AS, Cooper GA (1935) New formation names in the Michigan Devonian. J Wash Acad Sci 25:524–526

Warthin AS, Cooper GA (1943) Traverse rocks of Thunder bay region, Michigan. Bull Am Assoc Petrol Geol 27:571–595

Witzke BJ, Heckel PH (1988) Paleoclimatic indicators and inferred Devonian paleolatitudes of Euramerica. In: McMillan NJ, Embry AF, Glass DJ (eds) Devonian of the World. Canadian Society of Petroleum Geologists, Memoir 14, Calgary, pp 49–63

Wylie AS, Huntoon JE (2003) Log-curve amplitude slicing: visualization of log data and depositional trends in the Middle Devonian Traverse Group, Michigan basin, United States. Am Assoc Petrol Geol Bulletin 87:581–608

# Chapter 4
# Evolution in the Fossil Record

**Abstract** Our geological timescale based on stratigraphy and morphology for invertebrate evolution will be useful in calibrating local geological timescales and in estimating divergence times more reliably, especially in groups with poor fossil records. These morphological divergence times also provide an independent measure of the tempo and mode of morphological change. An accurate knowledge of divergence times can provide a clue to 'missing' fossils and thus enable the testing of various hypotheses of evolution in deep time.

## 4.1 Significance of Timescale in Evolutionary Studies

With the onset of innovations in the evolutionary theories, our knowledge of paleontological datasets and the fossil record has been improving (Stanley 1982; Smith 1984; Novacek 1992; Martin 1993; Benton and Storrs 1994; Gee 1996; Jablonski 2005). A detailed access to the geologic timescale is necessary for estimating rates and modes of morphological change in organisms and for interpreting macroevolutionary patterns and geographic variation. Traditionally, estimates of these times were obtained from the fossil record (Benton 1993). However, a more refined accuracy in timescales was obtained by determining the divergence-time estimates for mammalian orders and major lineages of vertebrates (Kumar and Hedges 1998; Polly 2008). Later, several studies were performed to determine the divergence times from lineages of invertebrates (Hunt and Chapman 2001, 2006, 2007, 2010; Bose 2012). This accurate determination of the geologic timescales provides us with a higher resolution fossil record and thus, paleontologists are slowly overcoming the challenges of examining the evolutionary patterns of extinct organisms by detecting the organisms' transformation stages preserved in the fossil record.

## 4.2 Evolution in Extinct Species

Extinct species preserve rich records of morphological change including major evolutionary novelties in geologic timescales. In this study, evolutionary patterns of extinct brachiopod fossil lineages have been investigated to better understand how they changed with time. The integration of various paleontological datasets in this study, reveal that the seemingly homogeneous group of brachiopods exhibit subtle but significant evolution in their morphology that is correlated with several kinds of ecological differences. The environmental changes influencing the morphology of these fossil lineages appear to be local in scope when compared with other geologically time equivalent fossil strata.

# References

Benton MJ (1993) The fossil record 2. Chapman and Hall, London

Benton MJ, Storrs GW (1994) Testing the quality of the fossil record: paleontological knowledge is improving. Geology 22:111–114

Bose R (2012) Biodiversity and evolutionary ecology of extinct organisms. Earth system sciences. Springer, New York

Gee H (1996) Before the backbone. Chapman and Hall, New York

Hunt G, Chapman RE (2001) Evaluating hypotheses of instar-grouping in arthropods: a maximum likelihood approach. Paleobiol 27(3):466–484

Hunt G (2006) Fitting and comparing models of phyletic evolution: random walks and beyond. Paleobiology 32:578–601

Hunt G (2007) The relative importance of directional change, random walks, and stasis in the evolution of fossil lineages. Proc Natl Acad Sci 104:18404–18408

Hunt G (2010) Evolution in fossil lineages: paleontology and the origin of species. Am Nat 176:S61–S76

Jablonski D (2005) Evolutionary innovations in the fossil record: the intersection of ecology, development, and macroevolution. J Exp Zool 304B:504–519

Kumar S, Hedges SB (1998) A molecular timescale for vertebrate evolution. Nature 392:917–920

Martin RD (1993) Primate origins: plugging the gaps. Nature 363:223–234

Novacek MJ (1992) Mammalian phylogeny: shaking the tree. Nature 356:121–125

Polly PD (2008) Adaptive Zones and the pinniped ankle: a 3D quantitative analysis of Carnivoran Tarsal evolution. In: Sargis E, Dagosto M (eds) Mammalian evolutionary morphology: a tribute to Frederick S. Szalay. Springer, Dordrecht, pp 165–194

Smith AB (1984) Systematics and the fossil record: documenting evolutionary patterns. Blackwell Science, Oxford, p 223

Stanley SM (1982) Macroevolution and the fossil record. Evolution 36:460–473

# About the Authors

## Rituparna Bose

Rituparna Bose (M.S., Ph.D.) is currently an adjunct Assistant Professor in the Department of Earth and Atmospheric Sciences and Department of Biological Sciences and Geology at the City University of New York. She serves on the editorial board of several scholarly journals—*Historical Biology: An International Journal of Palaeobiology* (Taylor and Francis), *Bulletins of American Paleontology* (Paleontological Research Institute, Cornell University), *SpringerPlus* (Springer), and *Geological Journal* (Wiley). She also serves as an Associate Editor-in-Chief for the *International Journal of Environmental Protection* and as an Associate Editor for the *Journal of Geography and Geology* at Canadian Center of Sciences and Education. She has also been invited to serve as an Editor for the journal *Acta Palaeontogica Sinica* (Chinese Academy of Sciences). Additionally, she reviews for several scholarly journals in the field of earth and environmental sciences.

During her doctoral career at Indiana University, Bloomington, she won major national awards like the Theodore Roosevelt Memorial Grant (American Museum of Natural History) and Schuchert and Dunbar Grant (Yale Peabody Museum of Natural History). Additionally, BP Global Energy Group funded her to present her research findings at North American Paleontology Convention (NAPC) by the prestigious NAPC Travel Award. She was also awarded the Indiana University Dissertation Year Research Fellowship which is given to the best doctoral students of the university.

Dr. Bose continues to pursue both her teaching and research career. Her research interests lie in applications of quantitative algorithms in evolutionary biology, paleoecology, paleoenvironment, and other applied biogeosciences. In addition, she teaches courses in Physical Geology, Environmental Geology, Earth System Science, Global Natural Disasters, and Historical Geology.

The foreword of this book has been written by Prof. Ashok Sahni, President of the Palaeontological Society of India and a fellow of the Indian National Science Academy.

R. Bose and A. J. Bartholomew, *Macroevolution in Deep Time*, SpringerBriefs in Evolutionary Biology, DOI: 10.1007/978-1-4614-6476-1, © The Author(s) 2013

# Alexander J. Bartholomew

Dr. Alexander Bartholomew (M.S., Ph.D.) is currently an Associate Professor in the Department of Geology at the State University of New York. Prof. Bartholomew obtained his Ph.D. from the University of Cincinnati, Ohio, United States. He is a Section Chair of the Paleontological Society of the United States. He also serves on the academic editorial board of the EvoS Journal: The Journal of the Evolutionary Studies Consortium. Additionally, he serves as a corresponding member to the UNESCO-IUGS Subcommission on Devonian Stratigraphy.

During his doctoral career at the University of Cincinnati, he won major national awards like the Theodore Roosevelt Memorial Grant (American Museum of Natural History) and the Graduate Research Grant from the Geological Society of America. Also while in graduate school, he was awarded the University of Cincinnati Graduate Assistant Excellence in Teaching Award in the McMicken College of Arts and Sciences.

Dr. Bartholomew continues to pursue both his teaching and research career with research interests centering upon paleoecological investigations of the Lower and Middle Devonian of eastern North America with a focus on biofacies stability through time and the timing of large-scale faunal change over broad geographic areas. At New Paltz, he teaches courses in Historical Geology, Paleontology, Stratigraphy and Sedimentology, Introduction to Speleology, as well as leading various field excursions to the western United States each year.

# Index

R. Bose and A. J. Bartholomew, *Macroevolution in Deep Time*, SpringerBriefs
in Evolutionary Biology, DOI: 10.1007/978-1-4614-6476-1,
© The Author(s) 2013